Symmetries and Conservation Laws in Particle Physics

An Introduction to Group Theory for
Particle Physicists

Symmetries and Conservation Laws in Particle Physics

An Introduction to Group Theory for Particle Physicists

Stephen Haywood

Rutherford Appleton Laboratory, UK

 World Scientific

NEW JERSEY · LONDON · SINGAPORE · BEIJING · SHANGHAI · HONG KONG · TAIPEI · CHENNAI

Published by

Imperial College Press
57 Shelton Street
Covent Garden
London WC2H 9HE

Distributed by

World Scientific Publishing Co. Pte. Ltd.
5 Toh Tuck Link, Singapore 596224
USA office: 27 Warren Street, Suite 401-402, Hackensack, NJ 07601
UK office: 57 Shelton Street, Covent Garden, London WC2H 9HE

Library of Congress Cataloging-in-Publication Data
Haywood, Stephen.
  Symmetries and conservation laws in particle physics : an introduction to group
theory for experimental particle physicists / Stephen Haywood.
    p. cm.
  ISBN 978-1-84816-659-2 (hardcover)
  ISBN 978-1-84816-703-2 (pbk.)
  1. Symmetry (Physics) 2. Conservation laws (Physics) 3. Group theory.
4. Transformation groups. 5. Particles (Nuclear physics) I. Title.
  QC174.17 .S9H39 2011
  539.7'25--dc22

                                                              2010032857

British Library Cataloguing-in-Publication Data

Printed in Singapore.

The heavens declare the glory of God;
the skies proclaim the work of his hands.
Day after day they pour forth speech;
night after night they display knowledge.
Psalm 19

Dedication

I dedicate this book to my parents, **Jim & Marion Haywood**, who encouraged me in my endeavours to become a physicist, and to **Jennifer** for all her love and support for the last 20 years.

The royalties from this book will go to the **Malawi Association for Christian Support** (MACS, http://www.malawimacs.org/) to support God's work in Malawi. The money raised will go some small way to helping people in one of the world's poorest countries – so please buy copies for all your friends and families.

Preface

In this book, we will explore the symmetries and conservation laws of the **Standard Model of Particle Physics** using the language of Group Theory. After an introduction to symmetries, conservations laws and groups, we will consider the properties of the groups $U(1)$, $SU(2)$ and $SU(3)$. These groups not only play a role in understanding the Standard Model, but provide a way of understanding hadrons through the **Quark Model** and an insight into physics **beyond the Standard Model**.

The key messages to extract from this book are:

- **Conserved quantities** arise from **Symmetries**.
- **Group Theory** provides the language for understanding how **particles combine** and the structure of the **Standard Model**.
- **Group Theory** provides insight, facilitating the **classification of particles** and the evaluation of their **interactions.**

This book is aimed at <u>first year PhD students in Experimental Particle Physics</u>. It is intended to fill the gap between some of the standard texts which have 2–3 pages on the subject, and much more advanced books aimed at theorists and mathematicians. Throughout this book, we will focus on **ideas** rather than developing formalism. We will draw on concepts and results from **Group Theory** without necessarily proving the assertions. There is a lot of mathematics behind what we need to know and it is not necessary for us to learn all of this. Instead, I will seek to make things plausible and will try to make as much connection as possible with concrete examples. Throughout the book, a classical approach to quantum mechanics, rather than field-theoretical approach, will be used. The reader should not lose sight of the fact that quantum mechanics is unnatural to us human beings. The subject is predicated on a set of rules which appear to describe observations of the microscopic world. At times, it will appear we get "something for nothing" – this is how the scientific process works. We make experimental observations; we try to encapsulate these in a mathematical theory, trying to include as much generality as possible. A good theory will make predictions beyond the original observations, and therefore be subject to further scientific investigation which will either support or invalidate the theory. But ultimately, everything comes back to experimental observation.

This book is based on a lecture course I have given in the University of London for the last nine years. It complements other courses on Quantum Field Theory and Gauge Theory. The course was originally conceived by **Paul Harrison** and then extended by **Alex Martin** – my thanks to them for all their groundwork. Also, my thanks to all the students over the years who have sharpened the material and provided some nice solutions to the associated problems. I am very grateful to **Ian Tomalin** for his very careful reading of the first draft and his many useful suggestions and to Chan Hong-Mo for valuable discussions. My thanks also to the team at ICP, in particular Kellye Curtis and Jackie Downs. I would like to thank the many friends who supported me in the tough times while I wrote this book, especially Rob & Sue, Peter & Fiona and Alan & Sarah. The funding for my time for this course has been provided by the Science and Technology Facilities Council (STFC), with the encouragement of the various directors of the **Particle Physics Department** at the **Rutherford Appleton Laboratory** (RAL).

Books I have found particularly useful in preparing the lecture course and writing this book include:
"Lie Algebras in Particle Physics" – H. Georgi
"Unitary Symmetry and Elementary Particles" – D. Lichtenberg

In Appendix A, hints and outline answers are provided for some of the problems – in the chapters, these problems are indicated with a "*". For teachers, a complete set of outline answers can be provided upon request to myself.

<u>Some typographical points:</u>

- Important terms are indicated by **bold** text; definitions are indicated by **<u>underlined bold</u>** text.
- Vectors are indicated usually by bold text, although sometimes 1D variables are used yet the extension to 3D is obvious.
- Hermitian conjugates are indicated by a superscript ^{H}.
- Throughout natural units with $\hbar = c = 1$ will be used.
- Abbreviations: "QM" = "Quantum Mechanics", "SM" = "Standard Model".
- I tend to use superscript letters a, b, c etc. to identify different particles where there are more than one, so as to distinguish between particle labels and powers.
- Usually, and unless stated otherwise, the convention of summation of repeated indices is used.
- Coupling constants in Lagrangians and covariant derivatives are frequently ignored to reduce clutter.

Stephen Haywood (RAL, Spring 2010)
Stephen.Haywood@stfc.ac.uk

Contents

Dedication   v

Preface   vii

1. Symmetries and Conservation Laws   **1**

  1.1 Introduction 1
  1.2 Symmetries and Transformations 1
      1.2.1 An illustration from non-relativistic mechanics 3
  1.3 Transformations in Quantum Mechanics 6
  1.4 Generators 9
      1.4.1 Exponentiation of operators 10
      1.4.2 Generators for translations 10
      1.4.3 Generators for rotations 12
  1.5 Symmetries in Quantum Mechanics 13
      1.5.1 Symmetries and conservation laws 14
      1.5.2 Symmetries without conserved observables 15
  1.6 Conservation Laws in Classical Mechanics 16
      1.6.1 The Lagrangian approach 16
      1.6.2 The Hamiltonian approach 17
      1.6.3 Connections with quantum mechanics 18
  1.7 Parity 19
  1.8 An Example 21
      1.8.1 Classical approach 21
      1.8.2 Quantum approach 22
  1.9 Problems 23

2. Introduction to Group Theory   **25**

  2.1 Introduction 25
  2.2 Groups 25

2.2.1 Example – a finite group 26
2.2.2 Subgroups and isomorphisms... 27
2.2.3 Example – a continuous group 28
2.3 Representations of Groups 29
2.3.1 Example – representations for rotations 29
2.3.2 Regular representations 30
2.3.3 Example – a regular representation for Z_3........................... 31
2.3.4 Irreducible representations 32
2.3.5 Example – reducible representations in parity transformations.......... 33
2.4 Lie Groups.. 35
2.4.1 More on generators 35
2.4.2 Lie algebra 37
2.4.3 The adjoint representation... 39
2.5 Unitary and Orthogonal Transformations... 41
2.5.1 Special unitary transformations... 42
2.5.2 Orthogonal transformations.. 43
2.5.3 More on the generators of U(n).. 44
2.6 Problems.. 45

3. The Unitary Group U(1)   **47**

3.1 Introduction.. 47
3.2 U(1).. 47
3.2.1 Global U(1) symmetry 48
3.2.2 Baryon and lepton number conservation..................................... 49
3.2.3 Local U(1) symmetry 51

4. The Special Unitary Group SU(2)   **53**

4.1 Introduction.. 53
4.2 2D Representations of the Generators 53
4.2.1 Quantum numbers in SU(2) 54
4.2.2 A 2D representation 55
4.2.3 An alternative 2D representation.. 56
4.3 A 3D Representation 58
4.4 Rotations 59
4.4.1 Comparison between spin and spatial rotations............................ 59
4.4.2 Combination of spin rotations 60
4.4.3 Rotation matrices 61
4.5 Gauge Transformations and the Adjoint Representation 65
4.6 Isospin 68

    4.6.1 Hadronic isospin 68

    4.6.2 Quark isospin 69

    4.6.3 Weak isospin 70

  4.7 Conjugate States 71

  4.8 Problems... 73

5. Combining Fermions   **75**

  5.1 Introduction.. 75

  5.2 Combining States 75

    5.2.1 Multiplets... 77

  5.3 Meson States 80

  5.4 Weights 81

    5.4.1 Weights in SU(2) 82

  5.5 Clebsch–Gordon Coefficients 84

6. The Special Unitary Group SU(3)   **87**

  6.1 Introduction.. 87

  6.2 The Gell-Mann Matrices 87

  6.3 Quantum Chromodynamics (QCD)..................................... 89

    6.3.1 The colour of hadronic states 90

    6.3.2 Gluons 93

    6.3.3 Colour factors.. 95

    6.3.4 Example – using colour factors 99

  6.4 Weights for SU(3) 100

  6.5 Quark Flavour in SU(3) 101

    6.5.1 Multi-particle states.. 106

  6.6 Young Tableaux – Constructing Multi-particle States 106

    6.6.1 Example – 2-particle states in SU(3)............................ 108

    6.6.2 Combining multiplets... 109

    6.6.3 Calculating multiplicities 111

    6.6.4 Examples – from SU(2) and SU(3) 114

  6.7 Problems.. 116

7. Hadron States   **117**

  7.1 Introduction.. 117

  7.2 Hadron States in SU(3)$_{\text{flavour}}$ 117

    7.2.1 Mesons 118

    7.2.2 Baryons 121

7.3 Hadron States in SU(6)$_{\text{flavour}\otimes\text{spin}}$... 126
7.4 Some Final Comments on Hadrons 129
7.5 Problems.. 129

8. The Standard Model and Beyond   **131**

8.1 Introduction 131
8.2 Consequences of Group Theory 131
8.3 The Standard Model 132
     8.3.1 Quantum numbers 135
8.4 The Higgs Mechanism... 136
8.5 Beyond the Standard Model 138
     8.5.1 SU(5).. 139
     8.5.2 Other symmetry groups.. 142
     8.5.3 SuperSymmetry.. 142

Appendix. Hints and Answers to Problems   143

Bibliography   149

Index   151

Chapter 1

Symmetries and Conservation Laws

1.1 Introduction

In this chapter, we will understand what **symmetries** are and how they are related to **transformations** in quantum mechanics.

The key message to extract from this chapter is:

- Symmetries give rise to conserved quantities.

1.2 Symmetries and Transformations

Systems are said to have <u>symmetry</u> if they are unchanged by a **transformation**. This symmetry is often due to an absence of an **absolute reference** and corresponds to the concept of **indistinguishablility**. It will turn out that symmetries are often associated with **conserved quantities**.

Transformations may be:

1. Active: Move the object – more physical.
2. Passive: Change the "description", e.g. change the coordinate frame – more mathematical. We will focus on these.

This is illustrated in Fig. 1.1.

We will consider two classes of transformation:

1. **Space-time transformations** (see Table 1.1):
   - Translations in (x, t)
   - Rotations and Lorentz boosts
   - Parity (reflections) in (x, t)

   where the first two sets of transformations constitute the so-called Poincaré transformations.
2. **Internal transformations**:
   - associated with the quantum numbers of a system

1

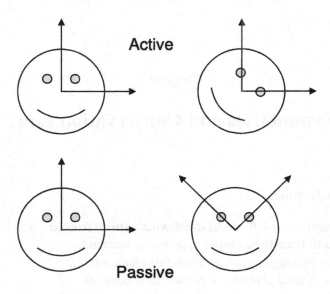

Fig. 1.1 Active and passive rotations.

Table 1.1 Space-time transformations.

Transformation	Description
Translations	$x \to x' = x - \Delta_x$ $t \to t' = t - \Delta_t$
Rotations (e.g. about z-axis)	$x \to x' = x\cos\theta_z + y\sin\theta_z$ $y \to y' = -x\sin\theta_z + y\cos\theta_z$
Lorentz boost (e.g. along x-axis)	$x \to x' = \gamma(x - \beta t)$ $t \to t' = \gamma(t - \beta x)$
Parity	$x \to x' = -x$
Time reversal	$t \to t' = -t$

For **physical laws** to be useful, they should exhibit a certain generality, especially under symmetry transformations. In particular, we should expect invariance of the laws to a change of the status of the observer – all observers should have the same laws, even if the evaluation of measurables is different. Put differently: the laws of physics, if applied correctly by different observers, should lead to the same expectations and observations. It is this principle which led to the formulation of Special Relativity.

1.2.1 *An illustration from non-relativistic mechanics*

Fig. 1.2 shows a mass m viewed by three observers. The boundary conditions for the mass are that at time $t = 0$, it is released from stationary, the spring having been stretched by an extension A:

$$x = W + L + A \quad \text{at } t = 0 \qquad (1.1)$$
$$\dot{x} = 0$$

Observer 1 measures things in a coordinate frame $\{x\}$, in which the *wall* is stationary. Newton's Laws of Motion give:

$$F = m\ddot{x}$$
$$\Rightarrow -k(x - (W + L)) = m\ddot{x} \qquad (1.2)$$

We can replace $x - (W+L)$ by ξ, the extension of the *spring*. This yields the equation for a simple harmonic oscillator:

$$-k\xi = m\ddot{\xi} \qquad (1.3)$$

Solving for ξ, and then x:

$$x = W + L + A\cos(\omega t) \quad \text{where} \quad \omega^2 = k/m \qquad (1.4)$$

Observer 2 measures in a coordinate frame $\{x'\}$. With respect to *Observer 1*, he is moving with a uniform velocity v:

$$x' = x - vt \qquad (1.5)$$

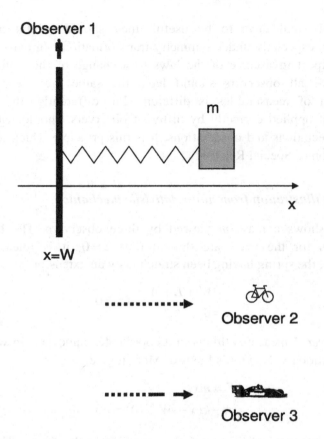

Fig. 1.2 A *mass m* is attached to a *wall* at $x = W$ by a *spring* of natural length L and spring constant k. The mass is observed by *Observers 1 & 2* in inertial frames, while *Observer 3* is in an accelerating frame.

As far as *Observer 2* is concerned:

$$-k(x' - (W' + L)) = m\ddot{x}' \qquad (1.6)$$

The expression for the force is unchanged. Again, we can denote the extension $x' - (W' + L)$ by ξ, where $W' = W - vt$. This is an appropriate thing to do because the 2nd derivative of W' vanishes, allowing us to replace the 2nd derivative of x with that of ξ. The solution is:

$$x' = W' + L + A\cos(\omega t) \quad \text{where} \quad \omega^2 = k/m \tag{1.7}$$

The frames of the two observers are equally good for the application of Newton's Laws; there is no way of distinguishing absolutely between the frames. This indistinguishability corresponds to a symmetry (or equality) between the two frames – associated with the transformation between the two.

The physical predictions of the two observers are identical: the amplitude and frequency are the same, although the description of the position of the *mass* is different.

However, **Observer 3** measures in a coordinate frame $\{x''\}$. With respect to *Observer 1*, he is accelerating with a uniform acceleration a:

$$x'' = x - \tfrac{1}{2}at^2 \tag{1.8}$$

Newton's Laws are only valid in inertial frames; so what happens if we try to apply them to an accelerating frame? As far as *Observer 3* is concerned:

$$-k(x'' - (W'' + L)) = m\ddot{x}'' \tag{1.9}$$

This time, if we replace $x'' - (W'' + L)$ by ξ, because the 2nd derivative of $W'' = W - \tfrac{1}{2}at^2$ does not vanish, the equation we obtain is slightly more complicated:

$$-k\xi = m(\ddot{\xi} - a) \tag{1.10}$$

and the solution is:

$$x'' = W'' + L + ma/k + (A - ma/k)\cos(\omega t) \tag{1.11}$$

This is a different solution – the amplitude is different. The reason is that the force was incorrectly attributed as $-k(x'' - (W'' + L))$. From *Observer 3*'s perspective, it appears that there is an additional force ma required to accelerate the *mass* – like a gravitational force.

The best way to solve for the motion of the mass, is to solve in frame $\{x\}$ and *then* transform. Why is there a lack of symmetry between the observers? We know that *Observer 3* (not *1* & *2*) is accelerating because he is forced into the back of his car seat.

The message is: one must be careful to understand the validity of given equation under a transformation.

1.3 Transformations in Quantum Mechanics

Consider a scalar wavefunction $\psi(x)$. If measured in a different coordinate frame $\{x'\}$, then the functional form of the wavefunction $\psi'(x')$ will change. However, at a given point in space, corresponding to x in one frame and x' in a second, because the amplitude (squared) of the wavefunction is a measurable quantity, the wavefunction should not depend on the coordinate system, and hence:

$$\psi'(x') = \psi(x) \qquad (1.12)$$

The intention is that x and x' correspond to the *same point* in space-time and the wavefunctions ψ' and ψ describe the *same event* – see Fig. 1.3.

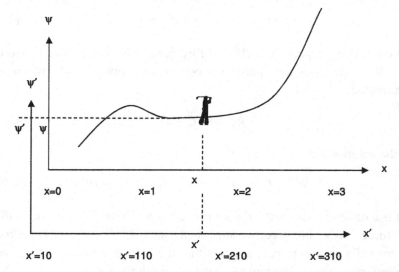

Fig. 1.3 In the first frame, a golfer striking a ball at x is described by $\psi(x)$. In the second frame, the event occurs at x' and is described by $\psi'(x')$. x and x' correspond to the same place, while $\psi(x)$ and $\psi'(x')$ correspond to the same event, namely the striking of the golf ball.

While ψ' and ψ describe the same event, their functional form will be different. If $x' = f(x)$, and the inverse transformation is f^{-1}, then $x = f^{-1}(x')$. Hence:

$$\psi'(x') = \psi(x) = \psi(f^{-1}(x')) \quad \text{so} \quad \psi'(\ ) = \psi(f^{-1}(\ )) \qquad (1.13)$$

For example, if $\psi(x) = x$ and $x' = \exp(x)$, then $\psi'(x') = \psi(x) = x = \log(x')$.

We assume that the new wavefunction can be derived from the old one by a transformation of the wavefunction itself:

$$\psi' = U\psi \qquad (1.14)$$

where U is an operator.

In general, ψ can be expressed as a linear superposition of base states $\{\phi_i\}$:

$$\psi = \sum c_i \phi_i \qquad (1.15)$$

In the new description, $\psi \rightarrow \psi'$, $\phi \rightarrow \phi'$, so:

$$\psi' = \sum c_i' \phi_i' \qquad (1.16)$$

But since ψ and ψ' correspond to the same states, as do ϕ and ϕ', then we would expect $c_i' = c_i$. Hence:

$$\psi' = \sum c_i \phi_i' \Rightarrow U(\sum c_i \phi_i) = \sum c_i U(\phi_i) \qquad (1.17)$$

This is the definition of a **linear operator**.

Note there are two distinct transformations:

    a. The transformation describing the change in "description" (coordinate frame): $x \rightarrow x' \equiv f(x)$.

    b. The corresponding transformation of the wavefunction: $\psi \rightarrow \psi' = U\psi$ – this results from the first transformation.

Furthermore, the overlap between any states ψ^a and ψ^b is an observable and should be independent of the description. Using the bra-ket notation for compactness: $\psi^a \leftrightarrow |a>$, then:

$$(\psi^a)' = U\psi^a \leftrightarrow |a'> = U|a> \qquad (1.18)$$

and:

$$(\psi^a)'^H = \psi^{aH} U^H \leftrightarrow <a'| = <a|U^H \qquad (1.19)$$

where ^{H} denotes the Hermitian conjugate. The overlap is:

$$<b'|a'> = <b|U^H U|a> \qquad (1.20)$$

and if this is to be equal to $<b|a>$ for all ψ^a and ψ^b, then:

$$U^H U = I \qquad (1.21)$$

This is the definition of a **unitary transformation**.

How do operators transform? If A is an operator, consider the corresponding observable:

$$<b|A|a> \equiv \int \psi^{bH} A \psi^a \qquad (1.22)$$

Since this an observable, and should be the same in all descriptions, we expect:

$$<b'|A'|a'> = <b|A|a> \qquad (1.23)$$

for all ψ^a and ψ^b. Hence:

$$<b|U^H A'U|a> = <b|A|a>$$
$$\Rightarrow U^H A'U = A \qquad (1.24)$$
$$\Rightarrow A' = UAU^H$$

If A happens to be invariant under the transformation, then:

$$A' = A$$
$$\Rightarrow UAU^H = A$$
$$\Rightarrow UA = AU \qquad (1.25)$$
$$\Rightarrow [A,U] = 0$$

i.e. A and U commute.

Lastly, if states $\{|i>\}$ form an orthonormal basis:

- $<j|i>=\delta_{ij}$
- All states $|a>$ in the vector space can be written as a linear superposition $|a>=\sum \alpha_i |i>$

Then the transformed states $|i'>=U|i>$ are also orthonormal:

$$<j'|i'>=<j|U^H U|i>=<j|I|i>=\delta_{ij} \qquad (1.26)$$

and since $|i'>$ is derived from $|i>$ etc., then if $i=j$, the states $|i'>$ and $|j'>$ must be equal, and hence the new base vectors are normal.

To demonstrate that they form a basis, consider representing the base vector $|i'>$ as a linear sum in the original basis:

$$|i'>=U|i>=\sum_j <j|U|i>|j> \qquad (1.27)$$

Since $<j|U|i>$ is just (the set of coefficients of) a unitary matrix, the matrix can be inverted, and then the indices i and j swapped:

$$|j>=\sum_i <i|U^H|j>|i'>$$
$$\Rightarrow |i>=\sum_j <j|U^H|i>|j'> \qquad (1.28)$$

Hence any state $|a>$ can be written as:

$$|a>=\sum_i \alpha_i |i>=\sum_i \alpha_i \sum_j <j|U^H|i>|j'>=\sum_j \beta_j |j'> \qquad (1.29)$$

which is the requirement for a basis.

1.4 Generators

Consider a transformation, $\psi \to \psi'=U\psi$. We assume that the unitary operator can be expressed as:

$$U=\exp(iaX) \qquad (1.30)$$

where a is real. But what is X ? Naively it is the "log" of U, but this is non-trivial, since we are dealing with operators and functions need to be expressed in terms of their power series. X is defined as the **generator** of the transformation.

1.4.1 *Exponentiation of operators*

We define the exponential of an operator in terms of a power series:

$$\exp(A) \equiv 1 + A + \tfrac{1}{2!}A^2 + \tfrac{1}{3!}A^3 + \ldots = \sum_{p=0}^{\infty} \tfrac{1}{p!}A^p \qquad (1.31)$$

where $A^0 = 1$ – the identity operator. It is easy to show that:

$$\exp(A)\exp(B) = \exp(A+B) \quad \text{if } [A,B]=0 \qquad (1.32)$$

However, if the operators do not commute, then this is not generally true. See Prob. 1.1.

If we consider the Hermitian conjugate of the exponential:

$$\exp(A)^H = (\sum_{p=0}^{\infty} \tfrac{1}{p!}A^p)^H = \sum_{p=0}^{\infty} \tfrac{1}{p!}(A^H)^p = \exp(A^H) \qquad (1.33)$$

If $U = \exp(iaX)$, then from Eq. (1.32), we can see $U^{-1} = \exp(-iaX)$. If U is unitary, then $U^{-1} = U^H$ hence:

$$\exp(-iaX) = (\exp(iaX))^H = \exp((iaX)^H) = \exp(-iaX^H) \qquad (1.34)$$

Equating terms order-by-order in a gives $X^H = X$, i.e. X is **Hermitian**. In QM, Hermitian operators are postulated to correspond to observables.

1.4.2 *Generators for translations*

Consider a translation in 1D in the x-direction:

$$x \to x' = x - \Delta_x \Rightarrow x = x' + \Delta_x \qquad (1.35)$$

So:

$$\psi'(x') = \psi(x) = \psi(x' + \Delta_x) \tag{1.36}$$

By Taylor expansion (and changing the dummy variable x' back to x for neatness):

$$\psi'(x) = \psi(x + \Delta_x)$$
$$= \psi(x) + \Delta_x \frac{\partial}{\partial x}\psi + \frac{1}{2!}\Delta_x^2 \frac{\partial^2}{\partial x^2}\psi + \frac{1}{3!}\Delta_x^3 \frac{\partial^3}{\partial x^3}\psi + \dots \tag{1.37}$$
$$= \exp(\Delta_x \frac{\partial}{\partial x})\psi$$

So we identify the corresponding unitary transformation as:

$$U = \exp(\Delta_x \frac{\partial}{\partial x}) \tag{1.38}$$

However, in QM, we associate the momentum operator with the spatial derivative:

$$p_x = -i\hbar \frac{\partial}{\partial x} \tag{1.39}$$

Hence:

$$U = \exp(i\Delta_x p_x / \hbar) \tag{1.40}$$

We often choose units so that $\hbar = 1$, and so $\hbar$ can be dropped. By reference to Eq. (1.30), we can identify the **generator of a translation** as the **momentum operator**.

This can be generalised to 3D: $x \to x' = x - \Delta \Rightarrow x = x' + \Delta$. Then:

$$\psi'(x) = \psi(x + \Delta)$$
$$= \psi(x) + \Delta \cdot \nabla\psi + \frac{1}{2!}(\Delta \cdot \nabla)^2\psi + \frac{1}{3!}(\Delta \cdot \nabla)^3\psi + \dots \tag{1.41}$$
$$= \exp(\Delta \cdot \nabla)\psi$$

where ∇ is the grad vector derivative. Identifying the 3D momentum operator as $p = -i\hbar\nabla$, we can write:

$$U = \exp(i\Delta \cdot p / \hbar) \tag{1.42}$$

Finally, extending what was done for spatial translations, we can identify the **generator of a time translation** as the **Hamiltonian operator**:

$$-i\frac{\partial}{\partial t} = -H / \hbar \tag{1.43}$$

1.4.3 *Generators for rotations*

Consider a rotation about the z-axis:

$$\begin{aligned} x \rightarrow x' &= x\cos\theta_z + y\sin\theta_z \\ y \rightarrow y' &= -x\sin\theta_z + y\cos\theta_z \end{aligned} \tag{1.44}$$

It proves to be much easier to consider infinitisimal rotations:

$$\begin{aligned} x \rightarrow x' &= x + y\theta_z \\ y \rightarrow y' &= -x\theta_z + y \end{aligned} \tag{1.45}$$

So:

$$\psi'(x', y') = \psi(x, y) = \psi(x' - y'\theta_z, y' + x'\theta_z) \tag{1.46}$$

and by Taylor expansion to first order and with a change of the dummy variables:

$$\begin{aligned} \psi'(x, y) &= \psi(x - y\theta_z, y + x\theta_z) \\ &= \psi(x, y) - y\theta_z\frac{\partial}{\partial x}\psi + x\theta_z\frac{\partial}{\partial y}\psi \\ &= \exp(\theta_z(x\frac{\partial}{\partial y} - y\frac{\partial}{\partial x}))\psi \end{aligned} \tag{1.47}$$

Recalling the definition of the orbital angular momentum operator:

$$L_z = (x \times p)_z = (xp_y - yp_x) = -i\hbar(x\frac{\partial}{\partial y} - y\frac{\partial}{\partial x}) \tag{1.48}$$

we can identify the **generator of a rotation** as the **angular momentum operator**.

Note: one should be careful about generalising this to 3D, since a rotation cannot be built up trivially of three rotations about the three orthogonal axes. The combination of three such rotations depends on the order. This will be manifested in QM if three operators are combined:

$$
\begin{aligned}
&\exp(i\theta_x L_x)\exp(i\theta_y L_y)\exp(i\theta_z L_z) \\
&\neq \exp(i(\theta_x L_x + \theta_y L_y + \theta_z L_z))
\end{aligned}
\tag{1.49}
$$

The reason is that $\exp(A)\exp(B) = \exp(A+B)$ only if A and B commute – which is not the case for the angular momentum operators (see Prob. 1.1). By contrast, the above *is* true for the momentum operators when generating 3D translations because of the commutation of the corresponding operators.

1.5 Symmetries in Quantum Mechanics

The "equation of motion" for a system in QM is given by:

$$
i\hbar\frac{\partial}{\partial t}\psi = H\psi
\tag{1.50}
$$

where $\psi = \psi(x,t)$ – a function of both space and time; while the Hamiltonian is usually considered to be time-independent: $H = H(x)$. Under a transformation, the wavefunction and Hamiltonian transform:

$$
\psi \to \psi' = U\psi \quad \text{and} \quad H \to H' = UHU^H
\tag{1.51}
$$

So in the new description:

$$
i\hbar\frac{\partial}{\partial t}\psi' = H'\psi'
\tag{1.52}
$$

By definition, the system is said to have a **symmetry** if the Hamiltonian is invariant, i.e. $H' = H$. Note: this is a symmetry of the Hamiltonian, not of the vector space (Hilbert space) of solutions $\{\psi\}$. It is H which defines the dynamics of the system, i.e. the interactions. Of course, the symmetry contained within H will be reflected in the individual solutions.

1.5.1 *Symmetries and conservation laws*

If a unitary transformation U is generated by X: $U = \exp(iaX)$. Then if H is invariant under the transformation U, then from Eq. (1.25):

$$[H,U]=0$$
$$\Rightarrow [H,\exp(iaX)]=0 \tag{1.53}$$
$$\Rightarrow [H,\sum \tfrac{1}{p!}(iaX)^p]=0$$

For this to be true for all orders of a:

$$\Rightarrow [H,X]=0 \tag{1.54}$$

Now consider the time variation of observables formed from X, such as $<b|X|a>$. Using the bra-ket notation and recalling that the Hamiltonian is Hermitian:

$$\tfrac{\partial}{\partial t}|a> = -\tfrac{i}{\hbar}H|a> \quad \text{and}$$

$$\tfrac{\partial}{\partial t}<b| = \tfrac{\partial}{\partial t}\{|b>\}^H = \{-\tfrac{i}{\hbar}H|b>\}^H = +\tfrac{i}{\hbar}<b|H \tag{1.55}$$

So:

$$\tfrac{d}{dt}<b|X|a>$$
$$= <b|\tfrac{\partial}{\partial t}X|a> -\tfrac{i}{\hbar}<b|X\,H|a> +\tfrac{i}{\hbar}<b|H\,X|a> \tag{1.56}$$
$$= <b|\tfrac{\partial}{\partial t}X|a> +\tfrac{i}{\hbar}<b|[H,X]|a>$$

So if X has no explicit time-dependence and $[H,X]=0$, then $<b|X|a>$ is constant in time.

To summarise: if the **Hamiltonian** of a system is **invariant** under a **unitary transformation** U generated by a Hermitian operator X, then there will be a **<u>conserved observable</u>** associated with X. Some of the transformations and their corresponding conserved observables in the case that the Hamiltonian is invariant are listed in Table 1.2.

Table 1.2 Transformations and their corresponding conserved observables.

Transformation	Conserved observable
Translation in space-time (x, t)	Momentum-energy (p, E)
Rotation in space	Orbital angular-momentum $L=x\times p$
Reflection in space	Parity
Gauge transformation	Charge

1.5.2 *Symmetries without conserved observables*

Previously it was shown Eq. (1.21) that transformation operators must be unitary. But strictly speaking, it is the modulus squared of amplitudes, $|<b\,|\,a>|^2$, not the amplitude, $<b\,|\,a>$, which should be invariant. So:

$$|UU^H| = 1 \tag{1.57}$$

It turns out that for **reflections in time** (time reversal), in some situations $UU^H = -1$ because the transformation is **antiunitary**, and consequently there is no corresponding conserved observable [Merzbacher].

Similarly, there are no conserved observables associated with **Lorentz boosts**. While these correspond to rotations in Minkowski space (x, it), they are not actually rotations in (x, t). Furthermore, the boost is not an isommetry – it does not preserve the volume element d^3x and hence the normalisation needs to be modified to preserve probability. In considering the invariance of the square of amplitudes, there is an implicit $\int d^3x$. So if we consider a Lorentz boost characterised by γ and where the wavefunctions are transformed by an operator L, then:

$$\psi \to \psi' = L\psi \quad \text{and} \quad d^3x \to d^3x' = \gamma d^3x \tag{1.58}$$

and the normalisation condition is:

$$\int \psi'^H \psi' d^3x' = \int \psi^H \psi d^3x = 1$$
$$\Rightarrow \int \psi^H L^H L\psi \gamma d^3x = 1 \tag{1.59}$$
$$\Rightarrow L^H L\gamma = 1$$

So L is not unitary.

In analogy with orbital angular momentum, one might have guessed a suitable generator would be proportional to $-tp_x + xH$, but the need to add a scale factor for the normalisation introduces an imaginary and therefore non-Hermitian term to the generator. Why the asymmetry with rotations? The Lorentz boost manifestly involves time (as well as spatial components) and what we are looking for are conserved observables which are constant in time.

1.6 Conservation Laws in Classical Mechanics

We have seen in QM that **symmetries** lead to **conservation laws**. Historically, some of the mathematical motivation for QM lies in the formulation of **classical mechanics**, on which we will touch briefly here. The classical formulations seem slightly perverse and maybe are a manifestation of the "real world" which is of course the quantum world. In the limit of large numbers of particles, the quantum world approximates to the description of classical mechanics. For these reasons, we shall not pursue this approach, but interested readers are referred to books on classical mechanics.

1.6.1 *The Lagrangian approach*

In classical mechanics, the starting point is the **Lagrangian**:

$$L(t,q,\dot{q}) = T - V \tag{1.60}$$

where T is the kinetic energy and V is the potential energy and q is a generalised coordinate. By minimising the **action**:

$$A = \int L(t,q,\dot{q})dt \tag{1.61}$$

one can derive the **Euler–Lagrange** equation of motion:

$$\frac{\partial L}{\partial q} - \frac{d}{dt}\frac{\partial L}{\partial \dot{q}} = 0 \tag{1.62}$$

The **canonical momentum** is defined:

$$p = \frac{\partial L}{\partial \dot{q}} \tag{1.63}$$

If the kinetic energy can be written as:

$$T = \tfrac{1}{2}\mu\dot{q}^2 \tag{1.64}$$

and V does not depend on $\dot{q}$, then:

$$p = \mu\dot{q} \tag{1.65}$$

– this looks like *mass* × *velocity*, although there is no reason for q to be a spatial coordinate, and hence $\dot{q}$ does not need to be velocity.

If V, and hence L, does not depend on q (kinetic energy does not usually depend on q), then:

$$\frac{\partial L}{\partial q} = 0 \tag{1.66}$$

and hence the Euler–Lagrange equation becomes:

$$\frac{d}{dt}p = 0 \tag{1.67}$$

which implies that p is constant. This situation corresponds to a uniform potential, i.e. not having derivatives with respect to q. Since we would tend to identify forces with the spatial derivatives of the potential, this corresponds to systems where, in the absence of external forces, the momentum is conserved.

1.6.2 *The Hamiltonian approach*

The **Hamiltonian** is constructed:

$$H(t, q, p) = p\dot{q} - L \tag{1.68}$$

The Hamiltonian equations of motion are:

$$\frac{\partial H}{\partial p} = \dot{q}, \quad \frac{\partial H}{\partial q} = -\dot{p} \quad \text{and} \quad \frac{\partial H}{\partial t} = -\frac{\partial L}{\partial t} \tag{1.69}$$

For many systems, this leads to a Hamiltonian which is equal to $T+V$, which we identify with the total energy of the system.

If V and hence L do not depend on time (kinetic energy does not usually depend on t):

$$\frac{\partial H}{\partial t} = 0 \tag{1.70}$$

and hence H is constant. We would tend to think of this as a situation where forces/potentials are not time-dependent and hence the total energy of the system is conserved.

1.6.3 *Connections with quantum mechanics*

Taking this further, it is possible to formulate the time variation of a quantity $Q(t, q, p)$:

$$\frac{dQ}{dt} = [Q, H] + \frac{\partial Q}{\partial t} \tag{1.71}$$

where the **Poisson Bracket** is defined by:

$$[Q, H] \equiv \frac{\partial Q}{\partial q}\frac{\partial H}{\partial p} - \frac{\partial Q}{\partial p}\frac{\partial H}{\partial q} \tag{1.72}$$

So for a simple system where:

$$H = \tfrac{1}{2}p^2 / \mu \tag{1.73}$$

$$\dot{q} = \frac{\partial H}{\partial p} = p/\mu \quad \text{and} \quad \dot{p} = -\frac{\partial H}{\partial q} = 0 \tag{1.74}$$

Then if Q does not depend on q and t, but only p:

$$\frac{dQ}{dt} = 0 \tag{1.75}$$

and Q is a constant in time.

As was done with QM, it is possible to identify **generators** of transformations and from the invariance of a Hamiltonian, deduce the presence of **conserved quantities**, such as momentum and angular momentum. Further, it is possible to identify analogies between the Poisson bracket formulation and the commutators of QM, as well as their corresponding Lie algebras. (Lie algebra will be discussed in the following chapter.) Again the interested reader is referred to text books on classical mechanics.

1.7 Parity

The **parity** transformation is a spatial reflection (in all 3 dimensions):

$$x \rightarrow x' = -x \tag{1.76}$$

We introduce the operator P which transforms the wavefunction:

$$\psi' = P\psi \tag{1.77}$$

For any isometry $d^3x \rightarrow d^3x' = d^3x$ (there is an implicit modulus) – since by definition an isometry does not alter the shape of an object. From the normalisation:

$$\int \psi'^H \psi' d^3x' = \int \psi^H P^H P\psi \, d^3x = 1$$
$$\Rightarrow P^H P = 1 \tag{1.78}$$

So we see the transformation is associated with a unitary transformation. Further, since:

$$\psi'(x') = \psi(x) = P\psi(x') \tag{1.79}$$

and using $x = -x'$, we have:

$$P\psi(x') = \psi(-x') \tag{1.80}$$

or replacing the dummy variable x' with x:

$$P\psi(x) = \psi(-x) \tag{1.81}$$

Therefore:

$$P^2\psi(x) = PP\psi(x) = P\psi(-x) = \psi(x)$$
$$\Rightarrow P^2 = I \tag{1.82}$$

So:

$$P^{-1} = P = P^H \tag{1.83}$$

Hence P is not only unitary but also Hermitian and corresponds to an observable.

If P has eigenstates $\{|\lambda>\}$ with eigenvalues $\{\lambda\}$, then:

$$P|\lambda> = \lambda|\lambda> \quad \text{and} \quad P^2|\lambda> = \lambda^2|\lambda> = |\lambda> \tag{1.84}$$

Hence:

$$\lambda^2 = 1$$
$$\Rightarrow \lambda = \pm 1 \tag{1.85}$$

So if parity is a symmetry of the Hamiltonian, then there exist states of well-defined parity (± 1) which will be conserved.

Observables are classified according to their transformation properties under parity transformations, as shown in Table 1.3.

Note: symmetries of the Hamiltonian must be verified experimentally. They may be postulated because they seem "sensible" and elegant, but this does not guarantee that they exist. For example, parity is *not* a symmetry of the weak interaction.

Table 1.3 Transformation properties of observables under parity.

Observable	Example	Transformation
Vector (or polar vector)	Spatial position Momentum	$x \rightarrow -x$ $p = -i\hbar\nabla \rightarrow -p$
Axial vector (or pseudovector)	Orbital angular momentum, also spin and total angular momentum	$L \equiv x \times p \rightarrow +L$
Scalar	Scalar product	$x \cdot x \rightarrow +x \cdot x$
Pseudoscalar	Helicity	$H \equiv \dfrac{J \cdot p}{|p|} \rightarrow -H$

1.8 An Example

Consider a Hamiltonian corresponding to two systems:
  a) Particle in a vacuum
  b) Particle subjected to a central force
The Hamiltonians are:

$$\text{a) } H_v = \tfrac{1}{2m} p^2 \quad \text{and} \quad \text{b) } H_{cf} = \tfrac{1}{2m} p^2 + V(r) \qquad (1.86)$$

We will investigate the effects of:
  i) Translations
  ii) Rotations (about the origin, on which the central force is centred)

1.8.1 *Classical approach*

Under a translation in 3D:

$$x \rightarrow x - \Delta \quad \text{and} \quad p = m\dot{x} \rightarrow p \qquad (1.87)$$

So p^2 is unchanged, but $r = \sqrt{x \cdot x}$ is changed.

Under a rotation with a rotation matrix R:

$$x \to Rx \quad \text{and} \quad p \to R p \tag{1.88}$$

So $p^2 = p^T p \to p^T R^T R p = p^T I p = p^T p$ is unchanged, as is r.
So under (i) translations, for the two systems:

    a)   The Hamiltonian is invariant and hence p is conserved.

    b)   The Hamiltonian is not invariant and hence p is not conserved.

And under (ii) rotations, for the two systems, the Hamiltonians are invariant and hence the angular momentum, L, is conserved.

1.8.2 *Quantum approach*

In the quantum approach, we consider the commutation of the Hamiltonian with the appropriate generators.

Translations are generated by the momentum operator, p. Clearly this commutes with the p^2 term in H_v. However, it does not commute with the term $V(r)$ in H_{cf}:

$$[V(r), p] \sim [V(r), \nabla] = V(r)\nabla - \nabla V(r) = -\nabla(V(r)) \tag{1.89}$$

In spherical coordinates, the radial component of the grad vector operator is the partial derivative with respect to r. So:

$$[V(r), p] \sim \hat{r} \frac{\partial V}{\partial r} \neq 0 \tag{1.90}$$

and therefore the momentum vector is not conserved in the presence of a central potential: a particle follows a conic-section path, not a straight line.

Rotations are generated by the angular momentum vector:

$$L \equiv x \times p \sim x \times \nabla \tag{1.91}$$

Firstly we consider the p^2 term in H_v:

$$[p^2, L] = [p^2, x \times p] = p^2(x) \times p + 2p(x) \times p \cdot p \tag{1.92}$$

where we have been a bit cavalier with the vectors and their indices, but it can all be followed through consistently, albeit with care. Now $\nabla^2(x) = 0$ and:

$$p(x) \times p \cdot p \sim \varepsilon_{abc} \partial_i (x_a) p_b p_i = \varepsilon_{abc} \delta_{ia} p_b p_i$$
$$= \varepsilon_{abc} p_b p_a = 0 \tag{1.93}$$

So $[p^2, L] = 0$ and the angular momentum is conserved in the absence of a force: the particle continues in a straight line, but its angular momentum about any origin is conserved.

Now we consider the potential term $V(r)$ in H_{cf}:

$$[V(r), L] = [V(r), x \times p] \sim x \times \nabla(V(r))$$
$$= x \times \hat{r} \frac{\partial V}{\partial r} = x \times \frac{x}{r} \frac{\partial V}{\partial r} = 0 \tag{1.94}$$

So the Hamiltonian and angular momentum commute and the angular momentum is conserved (provided it is evaluated with respect to the origin of the central force).

For all these examples, we see that the expectations of classical and quantum mechanics are the same.

1.9 Problems

Prob. 1.1* By considering the first few terms of the expansions, prove that:

$$\exp(A)\exp(B) \neq \exp(A + B)$$

in general, if A and B do not commute. However, show that the equality holds if A and B do commute (consider expansions to all orders).

Prob. 1.2* Find an expression for $\exp(i\alpha A)$ where:

$$A = \begin{pmatrix} 0 & 0 & 1 \\ 0 & 0 & 0 \\ 1 & 0 & 0 \end{pmatrix}$$

Prob. 1.3* If $[A, B] = B$, find an expression for:

$$f(\alpha) = \exp(i\alpha A) B \exp(-i\alpha A)$$

Chapter 2

Introduction to Group Theory

2.1 Introduction

In this chapter, we will learn what a **group** is. This will be illustrated by
the **unitary groups**.

The key messages to extract from this chapter are:
- Quantum mechanics is characterised by **vector spaces** of
  quantum numbers.
- We are interested in the properties of systems under
  transformations operating on the vector spaces.
- Transformations form **groups**.
- Group theory gives us the language and tools for understanding
  the properties.

2.2 Groups

A **group** consists of:
- Set of objects: $G = \{a, b, c \ldots\}$
- Binary operation for combining them: $\bullet$. Often the operation is
  "follows", and often one drops the operation sign.

It must satisfy:
- Closure: $\forall\, a, b \in G, a\bullet b \in G$
- Associativity: $\forall\, a, b, c \in G, (a\bullet b)\bullet c = a\bullet(b\bullet c)$
- Identity: $\exists\, e \in G$, s.t. $\forall\, a \in G, a\bullet e = a$
- Inverse: $\forall\, a \in G, \exists\, a^{-1} \in G$, s.t. $a\bullet a^{-1} = e$

In the above, we have only introduced the right-identity and the right-
inverse, but it is easy to prove that the identity is also a left-identity, and
the inverse is a left-inverse:

We introduce b, the right-inverse of a^{-1}: $a^{-1}\bullet b = e$

Then using the above rules:

$$e \bullet a = e \bullet (a \bullet e) = e \bullet (a \bullet (a^{-1} \bullet b)) = e \bullet ((a \bullet a^{-1}) \bullet b) = e \bullet (e \bullet b) = (e \bullet e) \bullet b = e \bullet b$$
$$= (a \bullet a^{-1}) \bullet b = a \bullet (a^{-1} \bullet b) = a \bullet e = a$$

And:

$$a^{-1} \bullet a = a^{-1} \bullet (a \bullet e) = a^{-1} \bullet (a \bullet (a^{-1} \bullet b)) = a^{-1} \bullet ((a \bullet a^{-1}) \bullet b) = a^{-1} \bullet (e \bullet b) =$$
$$(a^{-1} \bullet e) \bullet b = a^{-1} \bullet b = e$$

The operator need not be commutative, however, if the following criterion is satisfied:
- Commutativity: $\forall\ a, b \in G, a \bullet b = b \bullet a$

this group is said to be **<u>Abelian</u>**.

2.2.1 *Example – a finite group*

Consider a set of rotations:
- $R0$ = no rotation
- $R+$ = rotation of $+120°$
- $R-$ = rotation of $-120°$

Operation is "follows". Then the combination table is as shown in Table 2.1.

Table 2.1 Combination table for the cyclic group Z_3. (The significance of the shaded row is explained later.)

		2nd rotation		
"Follows"		$R0$	$R+$	$R-$
1st rotation	$R0$	$R0$	$R+$	$R-$
	$R+$	$R+$	$R-$	$R0$
	$R-$	$R-$	$R0$	$R+$

Let's check the group rules:

> Closed ✓
> Associative ✓
> Identity: $R0$
> Inverses: $(R0)^{-1} = R0, \quad (R+)^{-1} = R-, \quad (R-)^{-1} = R+$

This an example of the **cyclic group** of order 3: $\mathbf{Z_3}$. It is also **finite** since it has a finite number of elements.

2.2.2 Subgroups and isomorphisms

Consider permutations of a set of 3 objects $\{O1, O2, O3\}$. These could be a cat, a fish and a bird. These objects can be permuted between the locations where they start $\{L1, L2, L3\}$. There are $3! = 6$ permutations, which we will give labels $\{e, a, b, ...\}$ and describe in terms of which object is moved to which location. So for example, the permutation $(2,3,1)$ corresponds to moving object $O2$ to $L1$, $O3$ to $L2$ and $O1$ to $L3$, as in Fig. 2.1.

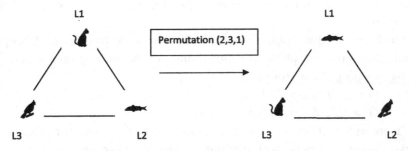

Fig. 2.1 Permutations of 3 objects between 3 locations.

So the set of permutations are:

> $e \equiv (1,2,3)$
> $a \equiv (2,3,1)$
> $b \equiv (3,1,2)$
> $x \equiv (1,3,2)$
> $y \equiv (3,2,1)$
> $z \equiv (2,1,3)$

Alternatively, one can think in terms of rotations $\{a, b\}$ and reflections $\{x, y, z\}$ of an equilateral triangle. The corresponding combination table is as shown in Table 2.2 and this corresponds to the **symmetry group** of order 3: S_3.

Table 2.2 Combination table for the symmetry group S_3.

	e	a	b	x	y	z
e	e	a	b	x	y	z
a	a	b	e	z	x	y
b	b	e	a	y	z	x
x	x	y	z	e	a	b
y	y	z	x	b	e	a
z	z	x	y	a	b	e

We see that the set $\{e, a, b\}$ forms a **subgroup** – a self-contained group. There are three other subgroups: $\{e, x\}$, $\{e, y\}$, $\{e, z\}$ – all examples of S_2.

Furthermore, the subgroup $\{e, a, b\}$ has the same form as Z_3. Groups which have different physical origins, and yet the same group structure are **isomorphic**. Two groups:

$G = \{a, b, c\}$ with $\bullet$
$G' = \{a', b', c'\}$ with $\blacksquare$

are isomorphic if there is a one-to-one correspondence or mapping between their members, such that the "products" also correspond to each other:

If $c = a \bullet b$ and $c' = a' \blacksquare b'$,
then for $\forall\ a, b, a', b'$:
$a \leftrightarrow a', b \leftrightarrow b'$ and $c \leftrightarrow c'$

Note: S_2 and Z_2 are isomorphic.

2.2.3 *Example – a continuous group*

Consider continuous rotations in 2D. A rotation by an angle α about the origin can be represented by a rotation matrix:

$$\begin{pmatrix} \cos\alpha & \sin\alpha \\ -\sin\alpha & \cos\alpha \end{pmatrix} \qquad (2.1)$$

These rotations can readily be shown to form a group. Because the parameter α is continuous and can take an infinite number of values, the group is said to be **continuous**. In 2D it is Abelian; but not in 3D. This is an example of a **special orthogonal** transformation in 2D: **SO(2)**.

Now consider multiplication of complex numbers, in particular by phasors: $\exp(i\alpha)$. This is an example of a **unitary** transformation in 1D: **U(1)**. It is obvious by consideration of the Argand plane that SO(2) and U(1) are isomorphic. These orthogonal and unitary transformations will be considered more later in this chapter.

2.3 Representations of Groups

The **representation** of a group $G = \{e, a, b, ...\}$ is a mapping D onto a set of **linear operators** acting on a vector space, such that:

$D(e) = I$ – the identity in the vector space

$D(a)D(b) = D(a\bullet b)$

In other words, the group "product" is mapped onto the multiplication operation in the vector space. This mapping is usually isomorphic. Often the operators are chosen to be matrices. Their form is not unique and depends explicitly on the vector space, in particular its dimensionality. It will transpire that we are more interested in the representations than the abstract groups. Note: every group has a trivial representation: $D(a) = 1$ for all group members a.

Also note: some authors use "representation" to refer to the vector space itself. In this book, we will use somewhat flexible nomenclature, depending on the circumstances.

2.3.1 *Example – representations for rotations*

U(1) has a natural representation in 1D, corresponding to phasors. However, this will also form a representation for SO(2).

SO(2) – rotations in 2D – have a natural representation as 2×2 matrices (the fundamental representation), but there are others, as shown in Table 2.3.

Table 2.3 Different representations of SO(2).

Vector space	Representation
2D space (x, y)	$\begin{pmatrix} \cos\alpha & \sin\alpha \\ -\sin\alpha & \cos\alpha \end{pmatrix}$
3D space (x, y, z)	$\begin{pmatrix} \cos\alpha & \sin\alpha & 0 \\ -\sin\alpha & \cos\alpha & 0 \\ 0 & 0 & 1 \end{pmatrix}$
Hilbert space for wave-functions	$\exp(i\alpha L_z) = \exp(\alpha(x\frac{\partial}{\partial y} - y\frac{\partial}{\partial x}))$
Complex numbers	$\exp(i\alpha)$

2.3.2 *Regular representations*

How do we form representations? Sometimes the physical concept behind the group will suggest a natural representation – as was obvious when we considered S_3 in terms of rotations and reflections. Another way is to form the so-called "regular representation" – this can be derived for any finite group using the technique described below.

We take the group elements to form a basis: {|e>, |a>, |b>, ...}. In vector notation:

$$|e> = \begin{pmatrix} 1 \\ 0 \\ 0 \\ ... \end{pmatrix}, \quad |a> = \begin{pmatrix} 0 \\ 1 \\ 0 \\ ... \end{pmatrix}, \text{ etc.} \tag{2.2}$$

We define the **regular representation** by the set of operators $\{D(a)\}$ such that:

$$D(a)|b> \equiv |a \bullet b> \qquad (2.3)$$

This is indeed a representation:

$$D(e)|a> = |e \bullet a> = |a>$$
$$\Rightarrow D(e) = I \qquad (2.4)$$

And:

$$D(a \bullet b)|c> = |(a \bullet b) \bullet c> = |a \bullet (b \bullet c)> = D(a)|b \bullet c>$$
$$= D(a)D(b)|c> \qquad (2.5)$$
$$\Rightarrow D(a \bullet b) = D(a)D(b)$$

The matrices can be constructed:

$$D(a)_{ij} \equiv <i|D(a)|j> = <i|a \bullet j> \qquad (2.6)$$

This corresponds to the projection of the vector corresponding to the product $|a \bullet j>$ onto $|i>$.

2.3.3 *Example – a regular representation for Z_3*

We recall the combination table for Z_3 given in Table 2.1. To find the operator $D(R+)$ corresponding to the rotation $R+$, we look at the elements $|R+ \bullet j>$ which correspond to those in the shaded row of the table – the column heading provides the "j" label.

So if we wish to find the column of the $D(R+)$ matrix corresponding to the "j" label for $R-$, we consider $|R+ \bullet R->$ – this is equal to $R0$. So the only "i" label with which this product will have a non-vanishing value is $R0$, resulting in an entry of "1" in the row for $R0$, but zeroes elsewhere. The process is illustrated in Fig. 2.2.

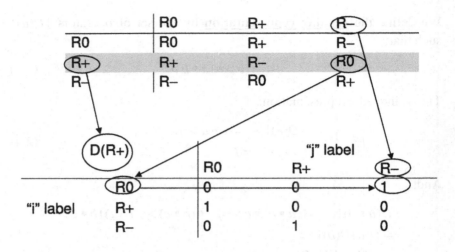

Fig. 2.2 Understanding the derivation of the regular representation. In this example, the matrix $D(R+)$ is obtained – see the text.

So the three matrices in the regular representation are:

$$D(R0) = \begin{pmatrix} 1 & 0 & 0 \\ 0 & 1 & 0 \\ 0 & 0 & 1 \end{pmatrix}, \quad D(R+) = \begin{pmatrix} 0 & 0 & 1 \\ 1 & 0 & 0 \\ 0 & 1 & 0 \end{pmatrix}$$

$$D(R-) = \begin{pmatrix} 0 & 1 & 0 \\ 0 & 0 & 1 \\ 1 & 0 & 0 \end{pmatrix}$$

(2.7)

These are matrices operating in 3D, corresponding to rotations:

$D(R0)$: identity
$D(R+)$: $x \rightarrow y, y \rightarrow z, z \rightarrow x$
$D(R-)$: $x \rightarrow z, y \rightarrow x, z \rightarrow y$

2.3.4 *Irreducible representations*

Representations are active on a **vector space**. If the action of all the operators on vectors $\{v\}$ in some **subspace** S is to produce vectors in the same subspace, then the representation is said to be **reducible**:

For $\forall\, v \in S$ and $\forall\, g \in G,\; D(g)\,v \in S$

For example in the regular representation of Z_3, there is an invariant subspace of all the points with $x = y = z$, i.e. all the points on this axis are left unchanged by the three rotations $D(R0)$, $D(R+)$ and $D(R-)$ of Eq. (2.7). Thinking of what Z_3 represents, this corresponds to (cyclic) permutations of 3 identical objects; think animals: consider 3 cats, or 3 fish. Care should be taken in always assuming the space is a real x-y-z space – it may not be!

It turns out that the subspaces will correspond to **multiplets** of particles: within the multiplet, the effect of operators from the group of transformations will be to create other particle states (or admixtures) within the same multiplet – see Section 5.2.1.

An **irreducible** representation is one which is not reducible. A **completely reducible representation** is one which can be broken down into a sum of independent irreducible representations.

2.3.5 *Example – reducible representations in parity transformations*

Parity (along with the identity operation) is an example of the group Z_2. The combination table is:

	e	P
e	e	P
P	P	e

The regular representation is [Georgi]:

$$D(e) = \begin{pmatrix} 1 & 0 \\ 0 & 1 \end{pmatrix}, \quad D(P) = \begin{pmatrix} 0 & 1 \\ 1 & 0 \end{pmatrix} \tag{2.8}$$

We can interpret $D(P)$ as a reflection in $x = y$. (One can envisage an infinite number of reflections in different lines, corresponding to alternative representations.)

To reduce the representation, we apply a similarity transformation S:

$$D(a) \rightarrow D'(a) = S^{-1}D(a)S \qquad (2.9)$$

This will lead to an alternative representation:

$$D'(a)D'(b) = S^{-1}D(a)S \cdot S^{-1}D(b)S$$
$$= S^{-1}D(a)D(b)S = S^{-1}D(a \bullet b)S = D'(a \bullet b) \qquad (2.10)$$

Here we make a 45° rotation with:

$$S = \frac{1}{\sqrt{2}}\begin{pmatrix} 1 & 1 \\ -1 & 1 \end{pmatrix} \qquad (2.11)$$

This gives:

$$D'(e) = \begin{pmatrix} 1 & 0 \\ 0 & 1 \end{pmatrix}, \quad D'(P) = \begin{pmatrix} 1 & 0 \\ 0 & -1 \end{pmatrix} \qquad (2.12)$$

So we can identify two independent subspaces:
    line $y = 0$, corresponding to $D'(e) = 1$, $D'(P) = +1$
    line $x = 0$, corresponding to $D'(e) = 1$, $D'(P) = -1$
The consequence of this is that we have succeeded in reducing the dimensionality of the vector spaces and the representations $D(e)$ and $D(P)$. See Section 5.2.1 for further insight.

If we have a Hamiltonian which is symmetric in the spatial coordinate x, then if we consider a parity transformation:

$$P : x \rightarrow x' = -x \qquad (2.13)$$

Then:

$$[H, P] = 0 \qquad (2.14)$$

implying that parity and energy can be simultaneous observables, and we can construct energy eigenfunctions which have well defined parity. In particular, the symmetric functions correspond to the representation with

$D(P) = +1$ and the antisymmetric functions correspond to the representation with $D(P) = -1$.

2.4 Lie Groups

Consider a **continuous group** with a finite number of parameters n – these could be labelled with a vector. For example:

    a)   Multiplication by a phasor $\exp(i\alpha)$ – 1 parameter

    b)   Rotation in 3D with parameters $\{\theta_x, \theta_y, \theta_z\}$ – 3 parameters

We can write the group members as $\{g(a)\}$, where a is a vector of n parameters. If the product is:

$$g(a'') = g(a) \bullet g(a') \tag{2.15}$$

then the group is a **Lie group** (pronounced "lee") if:

$$a'' = \Gamma(a, a') \tag{2.16}$$

where Γ is an analytic function. In practice, the kind of groups we encounter, which are characterised by parameters in an analytic way, will inevitably be Lie groups. The above examples are Lie groups:

    a)   For phasors, the function is simply the addition of the two phases.

    b)   For the rotations in 3D, the function is well-defined but more complicated.

2.4.1 *More on generators*

What is most interesting to us is to consider the effects of transformations on particle wavefunctions $\psi(x)$. Let us revisit what we looked at in Chapter 1, but with a bit more generality. Let us consider a group of transformations which act on the spatial coordinates x (implicitly a vector):

$$x \rightarrow x' = f(x, a) \tag{2.17}$$

where a is also implicitly a real vector. For example, this might correspond to a rotation or a Lorentz transformation. If we choose the identity transformation to be characterised by $a = 0$, and we consider an infinitesimal transformation δa, then:

$$x' = x + \delta x = f(x, \delta a) = f(x, 0) + \frac{\partial f}{\partial a} \delta a$$

$$\Rightarrow \delta x = \frac{\partial f}{\partial a} \delta a \qquad (2.18)$$

So:

$$\psi'(x') = \psi(x) = \psi(x' - \delta x) = \psi(x') - \frac{\partial \psi}{\partial x} \delta x \qquad (2.19)$$

Replacing the dummy variable x' by x:

$$\psi'(x) = \psi(x) - \frac{\partial \psi}{\partial x} \frac{\partial f}{\partial a} \delta a = (1 - \frac{\partial f}{\partial a} \frac{\partial}{\partial x} \delta a)\psi \qquad (2.20)$$

If we compare this with:

$$\psi' = U\psi = \exp(i \delta a X)\psi = (1 + iX \delta a)\psi \qquad (2.21)$$

we can identify:

$$X = i \frac{\partial f}{\partial a} \frac{\partial}{\partial x} \qquad (2.22)$$

as the **generator** for the transformation associated with a coordinate transformation (there is an implicit sum over the vector indices).

For each parameter (in the vector of parameters) there is one generator, so there are a total of n generators for the group of transformations.

Rather than considering coordinate transformations, one can consider a representation of the transformation acting in the Hilbert space:

$$\psi \rightarrow \psi' = D(\delta a)\psi = (1 + \frac{\partial D}{\partial a} \delta a)\psi \qquad (2.23)$$

where the generator is identified as:

$$X = -i \frac{\partial D}{\partial a} \qquad (2.24)$$

This is valid for transformations which do not operate on spatial coordinates, but on internal quantum numbers, for example spin or flavour indices.

As we saw in the last chapter, finite transformations can be built up by exponentiating the generators:

$$D(a_i) = \exp(i a_i X_i) \qquad (2.25)$$

where there is no summation over the index i. When combining transformations corresponding to different parameters or generators, care needs to be taken if the generators do not commute. With a suitable choice of parameters,[a] **Hermitian** generators can be chosen, leading to **Unitary** representations.

To summarise: often the easiest way to identify the generator of a transformation is to look at the infinitesimal transformation:

$$D(\delta a) \approx 1 + i X \delta a \qquad (2.26)$$

2.4.2 *Lie algebra*

Sophus Lie demonstrated that properties of Lie groups can be derived from consideration of elements which differ infinitesimally from the identity. We have seen when one considers one of the parameters, a_i, from the vector $a = (a_1, a_2, a_3, \ldots)$, the representation corresponding to the transformation is $D(a_i) = \exp(i a_i X_i)$ – no summation. It is therefore plausible (although not obvious) that the general representation corresponding to the vector of parameters a is:

$$D(a) = \exp(i \sum_i \Lambda_i(a) X_i) \qquad (2.27)$$

The sum is now shown explicitly and the functions $\Lambda_i(a)$ are not trivial as they were for a single parameter due to the potentially non-commuting nature of the generators. So any group member can be expressed by a representation which is the exponential of a linear combination of the

---

[a] I'd love to give you a reference for this, but since other books simply quote this, I will have to do the same.

generators. So if we consider the combination of two representations, this should result in a third:

$$D(a_1,0,0,...)D(0,a_2,0,...) = \exp(ia_1X_1)\exp(ia_2X_2)$$
$$= \exp(i\sum_i b_i X_i) \tag{2.28}$$

Expressing this a bit more generally, and using vector notation for $\{b_i\}$ and $\{X_i\}$:

$$\exp(ipX_p)\exp(iqX_q) = \exp(i\boldsymbol{b}\cdot\boldsymbol{X}) \tag{2.29}$$

where the scalar product is indicated explicitly by the "·". Expanding this to second order:

$$(1+ipX_p - \tfrac{1}{2}p^2X_p{}^2)(1+iqX_q - \tfrac{1}{2}q^2X_q{}^2)$$
$$= 1+i\boldsymbol{b}\cdot\boldsymbol{X} - \tfrac{1}{2}(\boldsymbol{b}\cdot\boldsymbol{X})^2 \tag{2.30}$$

Multiplying out:

$$1+ipX_p - \tfrac{1}{2}p^2X_p{}^2 +iqX_q - \tfrac{1}{2}q^2X_q{}^2 - pqX_pX_q$$
$$= 1+i\boldsymbol{b}\cdot\boldsymbol{X} - \tfrac{1}{2}(\boldsymbol{b}\cdot\boldsymbol{X})^2 \tag{2.31}$$

We want to solve implicitly for $\boldsymbol{b}\cdot\boldsymbol{X}$. Approximating to first order:

$$\boldsymbol{b}\cdot\boldsymbol{X} = pX_p + qX_q \tag{2.32}$$

Substituting this into the above:

$$1+ipX_p - \tfrac{1}{2}p^2X_p{}^2 +iqX_q - \tfrac{1}{2}q^2X_q{}^2 - pqX_pX_q$$
$$= 1+i\boldsymbol{b}\cdot\boldsymbol{X} - \tfrac{1}{2}(pX_p + qX_q)^2 \tag{2.33}$$

Expanding the last term and removing some of the obvious terms:

$$ipX_p + iqX_q - pqX_pX_q = i\boldsymbol{b}\cdot\boldsymbol{X} - \tfrac{1}{2}pX_pqX_q - \tfrac{1}{2}qX_qpX_p \tag{2.34}$$

Moving quadratic terms to the RHS:

$$ipX_p + iqX_q - i\boldsymbol{b} \cdot \boldsymbol{X} = \tfrac{1}{2} pq X_p X_q - \tfrac{1}{2} pq X_q X_p$$
$$= \tfrac{1}{2} pq [X_p, X_q] \tag{2.35}$$

So we see the commutator of the generators is a linear combination of the generators:

$$[X_a, X_b] = i\sum_c f_{abc} X_c \tag{2.36}$$

The exact definition varies slightly, sometimes "i" is missing. The coefficients f_{abc} are the **structure constants**. Although this is not a completely water-tight proof, it serves to illustrate the fundamentals and the can be shown to be valid for all orders.[b] By observation:

$$f_{bac} = -f_{abc} \tag{2.37}$$

Equation (2.36) defines the **Lie algebra** of the group. Through Eq. (2.35), the Lie algebra essentially determines the product of group members and hence the structure constants characterise the group. All representations of the group will have the same structure constants.

2.4.3 *The adjoint representation*

The adjoint representation for a group of transformations characterised by n parameters is a set of n $n \times n$ matrices $\{T^a : a = 1, ..., n\}$. The **adjoint representation** is derived from the **structure constants**:

$$T^a{}_{bc} = -if_{abc} \tag{2.38}$$

This expression defines the element (b, c) of the matrix T^a. Again, just as with the structure constants, there are some variations in the definitions with respect to the coefficient "$-i$".

---

[b] Again, I would love to give you a reference, but ...

Using the **Jacobi identity**:

$$[X_a,[X_b,X_c]]+[X_b,[X_c,X_a]]+[X_c,[X_a,X_b]]=0 \qquad (2.39)$$

and replacing the commutators with the structure constants:

$$[X_a,f_{bcp}X_p]+[X_b,f_{cap}X_p]+[X_c,f_{abp}X_p]=0 \qquad (2.40)$$

where there is implicit summation over p. Again replacing the commutators with the structure constants:

$$\Rightarrow f_{apq}f_{bcp}X_q + f_{bpq}f_{cap}X_q + f_{cpq}f_{abp}X_q =0 \qquad (2.41)$$

with summation over p and q. Assuming the generators form a basis:

$$\Rightarrow f_{apq}f_{bcp} + f_{bpq}f_{cap} + f_{cpq}f_{abp} =0 \qquad (2.42)$$

Swapping two indices, and using Eq. (2.37):

$$\Rightarrow f_{apq}f_{bcp} - f_{bpq}f_{acp} - f_{pcq}f_{abp} =0 \qquad (2.43)$$

Replacing some structure constants with components of the adjoint:

$$-T^a{}_{pq}T^b{}_{cp} +T^b{}_{pq}T^a{}_{cp} -iT^p{}_{cq}f_{abp} = 0$$
$$\Rightarrow T^a{}_{cp}T^b{}_{pq} -T^b{}_{cp}T^a{}_{pq} -if_{abp}T^p{}_{cq} = 0 \qquad (2.44)$$

Bearing in mind the summation over p and observing that we are seeing the (c, q) elements within a matrix equation:

$$\Rightarrow T^a T^b -T^b T^a -if_{abp}T^p = 0$$
$$\Rightarrow [T^a,T^b]= if_{abp}T^p \qquad (2.45)$$

So we see the adjoint also satisfies the Lie algebra and therefore provides an alternative set of generators of the representation.

It is turns out to be convenient [Georgi] to choose bases in which the adjoint satisfies:

$$\text{Tr}(T^a T^b) = \lambda \delta_{ab} \tag{2.46}$$

where λ is a scalar. This is achieved by a linear transformation of the generators. With this and Eq. (2.45), it is easy to show that:

$$f_{abc} = -i\lambda^{-1} \text{Tr}([T^a, T^b]T^c) \tag{2.47}$$

Since traces are unchanged by cyclically changing the order of matrices:

$$\begin{aligned}
\text{Tr}([T^a, T^b]T^c) &= \text{Tr}([T^a T^b T^c - T^b T^a T^c) \\
&= \text{Tr}([T^b T^c T^a - T^c T^b T^a) = \text{Tr}([T^b, T^c]T^a)
\end{aligned} \tag{2.48}$$

and similarly:

$$\text{Tr}([T^a, T^b]T^c) = -\text{Tr}([T^b, T^a]T^c) \tag{2.49}$$

So we see f_{abc} is unchanged by cyclic permutations and negated by anticyclic permutations – it is a completely antisymmetric tensor like the Levi–Civita tensor[c] ε_{abc}:

$$f_{abc} = f_{bca} = f_{cab} = -f_{acb} = -f_{cba} = -f_{bac} \tag{2.50}$$

2.5 Unitary and Orthogonal Transformations

The **<u>unitary group</u>** of order n, **U(n)**, is the group associated with $n \times n$ unitary matrices under matrix multiplication. The matrices operate on nD complex vectors. Unitary matrices U are defined by:

$$U^H U = U U^H = I \tag{2.51}$$

A complex $n \times n$ matrix has $2 \times n \times n$ parameters. Eq. (2.51) imposes $n \times n$ constraints. This is not totally trivial. Consider two distinct column vectors taken from the matrix U: a and b. The n diagonal elements of the matrix $U^H U$ must each satisfy an equation like $a^H a = 1$, of which there

---

[c] For SU(2), where the indices run from 1 to 3, the structure constants are the Levi–Civita tensor. This is not true for SU(n), where $n > 3$, as the indices may exceed 3.

are n of them. Since $a^H a$ is already real, they each provide one constraint. The off-diagonal terms are like $a^H b = 0$ and $b^H a = 0$. If the first is satisfied, the second is also true. However $a^H b$ is not necessarily real, and corresponds to two constraints: that both real and imaginary parts are zero. So these terms provide $n(n-1)/2$ pairs of constraints. So there are $n \times 1 + n(n-1)/2 \times 2 = n^2$ constraints. So there are n^2 free parameters and hence n^2 generators of U(n).

2.5.1 *Special unitary transformations*

Taking the determinants in the defining equation of a unitary transformation, Eq. (2.51):

$$\det(U^H U) = \det(I)$$
$$\Rightarrow \det(U^H)\det(U) = 1 \qquad (2.52)$$

Since:

$$U^H = U^{T*}$$
$$\Rightarrow \det(U^H) = \det(U^{T*}) = \det(U^T)^* = \det(U)^* \qquad (2.53)$$

Hence:

$$\det(U)^* \det(U) = 1$$
$$\Rightarrow |\det(U)|^2 = 1 \qquad (2.54)$$
$$\Rightarrow |\det(U)| = 1$$

The **<u>special unitary group</u>** is defined by $\det(U) = +1$. This additional constraint means **SU(n)** has $n^2 - 1$ generators.

Consider an infinitesimal unitary transformation corresponding to a parameter δa_i and with generators $\{X_i : i = 1, ..., n\}$. We will just consider one parameter at a time, and drop the index:

$$U(\delta a) = \exp(i\delta a X) = 1 + i\delta a X \qquad (2.55)$$

Where "1" is actually the identity operator, or "I". Remembering that δa is chosen to be real:

$$
\begin{aligned}
&U^H U = I \\
&\Rightarrow (1 - i\delta a X^H)(1 + i\delta a X) = I \\
&\Rightarrow -i\delta a X^H + i\delta a X = 0 \\
&\Rightarrow X^H = X
\end{aligned}
\tag{2.56}
$$

So the generators are Hermitian, as was seen in Section 1.4.1.

It is easy to show that for a matrix ε whose components are infinitesimally small:

$$
\det(1 + \varepsilon) = 1 + \mathrm{Tr}(\varepsilon) \tag{2.57}
$$

Therefore for SU(n), only keeping terms of order δa:

$$
\begin{aligned}
&\det(U) = 1 \\
&\Rightarrow \det(1 + i\delta a X) = 1 \\
&\Rightarrow 1 + i\delta a\, \mathrm{Tr}(X) = 1 \\
&\Rightarrow \mathrm{Tr}(X) = 0
\end{aligned}
\tag{2.58}
$$

So the generators of SU(n) are both Hermitian and traceless.

2.5.2 *Orthogonal transformations*

Important subgroups of U(n) and SU(n) are the **orthogonal groups O(n)** and **SO(n)** respectively. These are the real number versions of the unitary transformations and correspond to **isometries** in nD – an isometry does not change the shape of an object, e.g. rotations and reflections. Orthogonal matrices O are defined by:

$$
O^T O = O O^T = I \tag{2.59}
$$

In analogy with the discussion for the unitary matrices, there are n^2 parameters. The diagonal terms are like $a^T a = 1$ and provide one constraint each. The off-diagonal terms are like $a^T b = b^T a = 0$ and again provide one constraint each. So there are $n \times 1 + n(n-1)/2 \times 1 = n(n+1)/2$ constraints. This gives $n(n-1)/2$ free parameters.

As before, we can show:

$$\begin{aligned} |\det(O)| &= 1 \\ \Rightarrow \det(O) &= \pm 1 \end{aligned} \tag{2.60}$$

This time, the determinant can be restricted to ± 1, since the matrices are real.

SO(n) is derived from O(n) with the constraint: $\det(O) = +1$. These transformations correspond to rotations, while those with $\det(O) = -1$ correspond to reflections (combined with rotations). This further constraint does not change the number of free parameters, since it relates to sign changes of the parameters. For example in SO(2):

$$O = \begin{pmatrix} \cos\theta & \sin\theta \\ -s \cdot \sin\theta & s \cdot \cos\theta \end{pmatrix} \tag{2.61}$$

where $s = +1$ leads to $\det(O) = +1$ and $s = -1$ leads to $\det(O) = -1$. For both, there is still one free parameter, θ.

2.5.3 *More on the generators of U(n)*

There are different representations[d] (infinitely many) of the generators of a group. For U(n), one can choose a representation of the generators which includes the unit matrix. (Recall we are talking about the set of generators – not the group members – which must include the identity.) However Tr(I) = n, so I cannot be a generator for SU(n) ... although of course I is a member of SU(n). The generator I corresponds to a phase change:

---

[d] In this section, "representations" means "expressions" or "forms", as distinct from the definition introduced in Section 2.3 for groups. The generators do not necessarily form a group. For example, in Chapter 4, we will see that the Pauli spin matrices are suitable generators for SU(2), however they do not form a group under multiplication (they do not include the identity, for a starter). In the same chapter, we will see different ways of writing the SU(2) generators.

$$U = \exp(i\alpha I) = \exp(i\alpha) \qquad (2.62)$$

where the final expression is implicitly multiplied by the $n \times n$ unit matrix. This matrix corresponds to the group U(1).

So U(n) is made up of SU(n) and a U(1) – U(n) is isomorphic to the combination of the two and is written as the "product":

$$U(n) = SU(n) \otimes U(1) \qquad (2.63)$$

The complete set of U(n) operators can be constructed from the combination (multiplication) of the set of operators found in SU(n) and those found in U(1).

2.6 Problems

Prob. 2.1* Consider which of the following are groups:
   a)   Integers under Addition
   b)   Integers under Subtraction
   c)   Integers under Multiplication
   d)   Reals under Multiplication

Prob. 2.2 Demonstrate that there is only one group combination table for 3 distinct objects, i.e. all groups for 3 objects have the same form (are isomorphic) to Z_3.

Prob. 2.3* Show that the set of Lorentz transformations:

$$g(\beta) \quad \begin{cases} x' = \gamma(x - \beta t) \\ t' = \gamma(t - \beta x) \quad |\beta| < 1 \\ \gamma = 1/\sqrt{1 - \beta^2} \end{cases}$$

form an Abelian Lie group under the operation "follows".

Prob. 2.4 Show that U(n) and SU(n) are groups.

Prob. 2.5* Consider rotations in 3D about the x-, y- and z-axes – SO(3). Identify generators appropriate to

    a)   Scalar wavefunctions $\psi(x)$

    b)   Real vectors in 3D space

In both cases, find the structure constants.

Prob. 2.6* For the generators $\{L_x, L_y, L_z\}$ in part (b) of Prob. 2.5, find the simultaneous eigenvectors of L^2 and L_z.

Prob. 2.7* Find the adjoint matrices for the generators in part (b) of Prob. 2.5. In this case, it is obvious that they satisfy the Lie algebra.

Chapter 3

The Unitary Group U(1)

3.1 Introduction

In this chapter, we will consider the unitary group **U(1)** and have a first look at **gauge symmetries**.

The key messages to extract from this chapter are:

- The group U(1) is associated with the conservation of **baryon and lepton numbers**.
- Local U(1) symmetry underlies **QED** (quantum electro-dynamics).

3.2 U(1)

<u>U(1)</u> is the group of **unitary transformations** acting on complex 1D vectors, i.e. complex numbers. This corresponds to a phase transformation and a suitable generator is the unit operator 1.

However, rather than a vector space consisting of complex numbers, we can choose a different vector space, namely one consisting of the set of particle wavefunctions in Hilbert space and for which we choose to operate on a single measurable quantity characterised by some quantum number. In this case, the generator is not the unit operator, but the appropriate quantum operator.

Let us consider the quantity of electric charge. We will consider groups of transformations of the form:

$$U = \exp(i\alpha Q) \tag{3.1}$$

The generator is Q, the charge operator, and this acts on the 1D vector space identified with the electric charge. We choose to consider transformations which are either:

  <u>**Global**</u>: α does not depend on spatial position x, or
  <u>**Local**</u>: α does depend on x

3.2.1 *Global U(1) symmetry*

Consider the example of electric charge, with an operator Q. If the (interaction) Hamiltonian H contains a symmetry with respect to the transformation of Eq. (3.1), then it commutes with Q, and we know the charge of a state will be conserved. Previously we considered the time variation of an observable. Alternatively, in Chapter 1 we found that the Hamiltonian is the generator for time translation. So the time evolution of a state ψ of some well defined charge, q, is given by:

$$\psi \rightarrow \psi' = \exp(-itH)\psi \qquad (3.2)$$

Consider the charge of the final state, determined by the charge operator:

$$\begin{aligned} Q\psi' &= Q\exp(-itH)\psi = \exp(-itH)Q\psi \\ &= \exp(-itH)q\psi = q\exp(-itH)\psi = q\psi' \end{aligned} \qquad (3.3)$$

We see the final state is also a state of well-defined charge, being the same as the original one.

By construction, the Standard Model Lagrangian is invariant under the global transformation associated with electric charge. It consists of terms like:

$$L = \psi^a \psi^b \psi^c ... \qquad (3.4)$$

where, to ensure charge conservation, the sum of the charges of the corresponding particles is:

$$q^a + q^b + q^c + ... = 0 \qquad (3.5)$$

Under a U(1) transformation:

$$\psi \to \psi' = \exp(i\alpha Q)\psi \qquad (3.6)$$

which for a single particle gives:

$$\psi' = \exp(i\alpha q)\psi \qquad (3.7)$$

For multiple particles, the charge operator is made up of the sum of the operators for the individual particles:

$$Q = Q^a + Q^b + Q^c + ... \qquad (3.8)$$

Since the operators are all independent, they commute; hence:

$$\exp(i\alpha Q) = \exp(i\alpha Q^a)\exp(i\alpha Q^b)\exp(i\alpha Q^c)... \qquad (3.9)$$

and:

$$
\begin{aligned}
L' &= \exp(i\alpha Q)\psi^a\psi^b\psi^c... \\
&= \exp(i\alpha Q^a)\psi^a \exp(i\alpha Q^b)\psi^b \exp(i\alpha Q^c)\psi^c... \\
&= \exp(i\alpha q^a)\psi^a \exp(i\alpha q^b)\psi^b \exp(i\alpha q^c)\psi^c... \\
&= \exp(i\alpha(q^a + q^b + q^c + ...))\psi^a\psi^b\psi^c... \\
&= \exp(i\alpha 0)\psi^a\psi^b\psi^c... = \psi^a\psi^b\psi^c... = L
\end{aligned}
\qquad (3.10)
$$

We have not got something for nothing ... this is all by construction. While this seems pretty trivial, things will become more interesting when we consider the local transformations.

3.2.2 *Baryon and lepton number conservation*

By observation, **baryon (B)** and **lepton (L) numbers** are conserved in interactions. Therefore the Standard Model is constructed to ensure this. Baryon and lepton numbers are additive, just like charge. The numbers for the Standard Model particles are given in Table 3.1.

Table 3.1 Baryon and lepton numbers of Standard Model particles.

Particle	Baryon Number (B)	Lepton Number (L)
quark	$+^1/_3$	0
anti-quark	$-^1/_3$	0
lepton, neutrino	0	+1
anti-lepton, anti-neutrino	0	−1
Others (gauge bosons)	0	0

By construction the baryons, being made from three quarks, have baryon number +1.

The postulated conservation law means that within the Standard Model, a quark cannot be destroyed, except by annihilation with an anti-quark.

Until the recent observation of **neutrino oscillations**, it was suggested that the different flavours of lepton had their own separate conservation laws. The existence of right-handed neutrinos allows the possibility that neutrinos have mass and that the corresponding mass eigenstates are no longer eigenstates of the weak interaction, allowing mixing between the different flavours of neutrinos, as found in the quark sector.

In models derived with alternative symmetries, which are constructed in attempt to provide a **Grand Unified Theory**, baryon number and lepton number may not be conserved. In many models, such as SU(5) (see Chapter 8), B and L are not conserved but $B-L$ is. The possible decay of a proton $p \rightarrow e^+\pi^0$ mediated by a new gauge boson, X, is illustrated in Fig. 3.1. The proton has $B=+1$, $L=0$ and hence $B-L=+1$. The final state has $B=0$, $L=-1$ and hence $B-L=+1$ and we see $B-L$ is conserved. However, simple SU(5) models have been ruled out by the non-observation of proton decay (see [PDG] for a review of the measurements).

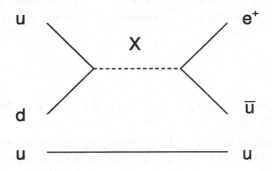

Fig. 3.1 Feynman diagram for proton decay in SU(5).

3.2.3 *Local U(1) symmetry*

Now we consider **local U(1) transformations**:

$$U = \exp(i\alpha(x)Q) \tag{3.11}$$

This is an example of a **gauge transformation**.[a] What are the consequences for the Lagrangian representing a fermion of charge q? The appropriate part of the Lagrangian is formulated as:

$$L \sim \overline{\psi}\partial\!\!\!/\psi \quad \text{where} \quad \partial\!\!\!/ \equiv \gamma_\mu\partial^\mu = \gamma_\mu\frac{\partial}{\partial x_\mu} \tag{3.12}$$

where there is an implicit sum over the Lorentz indices, μ. Under the local gauge transformation $\psi \rightarrow \psi' = U\psi$, the Lagrangian becomes:

$$L \rightarrow L' \sim L + \overline{\psi}(iq\partial\!\!\!/\alpha)\psi \neq L \tag{3.13}$$

because now the derivative of α does not vanish. If we wish to ensure invariance, we must replace ∂^μ by the covariant derivate:

$$D^\mu = \partial^\mu - iqA^\mu \tag{3.14}$$

---

[a] "Gauge" initially referred to "size", as in "railway track gauge". It is a misnomer for what is actually a phase transformation.

This requires the introduction of a field A^μ, and to ensure the invariance under the gauge transformation (gauge invariance), A^μ must transform according to:

$$A^\mu \rightarrow A^\mu + \frac{1}{q}\partial^\mu \alpha \qquad (3.15)$$

A^μ is identified as the field describing the photon – the quantum of electromagnetic interactions.

So by assuming a local U(1) symmetry (gauge invariance) associated with the electric charge, we have motivated the existence of the photon and subsequently electromagnetic interactions. Historically, it was the other way round. Electromagnetism was observed in macroscopic experiments, and Maxwell's equations were constructed to describe the fields. Subsequently it was noticed that they possessed a symmetry which was denoted by "gauge invariance". So are these mathematical "games" of any physical benefit? By postulating local SU(n) gauge symmetries associated with quantum numbers (observed or hypothesised), it is possible to predict vector fields corresponding to possible interactions. Only experiments can confirm whether these interactions are realised in nature. It was precisely this mode of thinking associated with SU(2) which led to the predictions of the $W^\pm$ and Z bosons ... but encouraged by a great deal of experimental input.

The above provides a very brief overview of gauge theory. Gauge theories are well described in many of the standard textbooks and will be touched on again in the next chapter.

Chapter 4

The Special Unitary Group SU(2)

4.1 Introduction

In this chapter, we will look at **different representations** of SU(2). We will consider **rotations** and have a brief look at **rotation matrices**. We will revisit **gauge transformations** and the **adjoint representation**. We will look at **isospin**, both in **hadronic systems** and also in the **Weak Interaction**. Finally, we will see how to derive the **conjugate states** corresponding to anti-particles.

The key messages to extract from this chapter are:

- SU(2) describes **spin angular momentum**.
- SU(2) is isomorphic to the description of **orbital angular momentum**, associated with **rotations** and the group **SO(3)**.
- SU(2) also describes **isospin** – of relevance to nucleons, light quarks and the Weak Interaction.

4.2 2D Representations of the Generators

<u>SU(2)</u> corresponds to the set of special unitary transformations which act on complex 2D vectors and the operation of matrix multiplication. The natural representation is that of 2×2 matrices acting on 2D vectors; nevertheless there are other representations, in particular in higher dimensions, as we shall see. There are 2^2-1 parameters, hence 3 generators, let us call them J_1, J_2, J_3. As seen in Chapter 2, the generators are traceless and Hermitian. It is easy to show that the matrices have the form:

$$\begin{pmatrix} a & b^* \\ b & -a \end{pmatrix} \qquad (4.1)$$

where a is real. There are three parameters: a, Re(b), Im(b). A suitable (but not unique) set of generators is provided by the Pauli spin matrices:

$$J_i = \tfrac{1}{2}\sigma_i \qquad (4.2)$$

where:

$$\sigma_1 = \begin{pmatrix} 0 & 1 \\ 1 & 0 \end{pmatrix}, \quad \sigma_2 = \begin{pmatrix} 0 & -i \\ i & 0 \end{pmatrix}, \quad \sigma_3 = \begin{pmatrix} 1 & 0 \\ 0 & -1 \end{pmatrix} \qquad (4.3)$$

These matrices have the properties that:

$$\sigma_i \sigma_j = \delta_{ij} + i\varepsilon_{ijk}\sigma_k \qquad (4.4)$$

The Lie algebra is:

$$[J_i, J_j] = i\varepsilon_{ijk}J_k \qquad (4.5)$$

There is an implicit sum over k, but in this case it is trivial: the commutator just gives the "other" generator, for example:

$$[J_1, J_2] = iJ_3 \qquad (4.6)$$

So we identify the structure constants for SU(2) as being the Levi–Civita tensor, ε_{ijk}.

4.2.1 *Quantum numbers in SU(2)*

A **<u>Casimir operator</u>** is one which commutes with all of the generators. In SU(2) there is just one Casimir:

$$J^2 = J_1^2 + J_2^2 + J_3^2 \qquad (4.7)$$

Since $[J^2, J_3] = 0$, J^2 and J_3 can have simultaneous observables and can provide suitable eigenvalues by which states can be labelled.

We can define **raising & lowering operators**:

$$J_\pm = J_1 \pm iJ_2 \tag{4.8}$$

It is easy to show that $[J_3, J_\pm] = \pm J_\pm$. We define **eigenstates** $| j\, m >$ with their corresponding **eigenvalues**:

$$J^2 \, | j m > = j(j+1) | j m > \tag{4.9}$$

$$J_3 \, | j m > = m | j m > \tag{4.10}$$

We recall from undergraduate courses on angular momentum in QM, J_+ and J_- generate different states within a multiplet:

$$\begin{aligned} J_3 J_\pm \, | j m > &= (m \pm 1) J_\pm \, | j m > \\ J^2 J_\pm \, | j m > &= j(j+1) J_\pm \, | j m > \end{aligned} \tag{4.11}$$

and:

$$J_\pm \, | j m > = \sqrt{(j \mp m)(j \pm m + 1)} \, | j m \pm 1 > \tag{4.12}$$

4.2.2 A 2D representation

In the previous sections, we started in 2D to motivate the SU(2) group structure (which is defined in 2D in the first place). Then we derived the **Lie algebra** and identified a scheme for labelling eigenstates of the generators. Further, we recalled results from angular momentum theory, although we have not explicitly restricted ourselves to angular momentum.

We can form a 2D representation by choosing two orthogonal states as base vectors:

$$\begin{pmatrix} 1 \\ 0 \end{pmatrix} \equiv | \tfrac{1}{2} +\tfrac{1}{2} > \quad \text{and} \quad \begin{pmatrix} 0 \\ 1 \end{pmatrix} \equiv | \tfrac{1}{2} -\tfrac{1}{2} > \tag{4.13}$$

By inspection:

$$J_3 = \begin{pmatrix} \frac{1}{2} & 0 \\ 0 & -\frac{1}{2} \end{pmatrix} \tag{4.14}$$

To find J_+, we consider:

$$J_+ \mid \tfrac{1}{2} + \tfrac{1}{2} > = 0 \quad \text{and} \quad J_+ \mid \tfrac{1}{2} - \tfrac{1}{2} > = \mid \tfrac{1}{2} + \tfrac{1}{2} > \tag{4.15}$$

$$\Rightarrow J_+ = \begin{pmatrix} 0 & 1 \\ 0 & 0 \end{pmatrix} \tag{4.16}$$

We can do the same for J_-. Alternatively, recalling the definition of $J_\pm$ and the fact that the generators are Hermitian:

$$J_- = J_1 - iJ_2 = (J_1 + iJ_2)^H = J_+{}^H = \begin{pmatrix} 0 & 0 \\ 1 & 0 \end{pmatrix} \tag{4.17}$$

Hence:

$$J_1 = \tfrac{1}{2}(J_+ + J_-) = \begin{pmatrix} 0 & \frac{1}{2} \\ \frac{1}{2} & 0 \end{pmatrix}$$

$$\text{and} \tag{4.18}$$

$$J_2 = \tfrac{1}{2i}(J_+ - J_-) = \begin{pmatrix} 0 & \frac{-i}{2} \\ \frac{i}{2} & 0 \end{pmatrix}$$

and we recover $J_i = \sigma_i/2$.

4.2.3 *An alternative 2D representation*

The previous result arose from a particular choice (the most natural) of base states. If we make another choice such that:

$$\begin{pmatrix} 1 \\ 0 \end{pmatrix} \equiv \tfrac{1}{\sqrt{2}}(\mid \tfrac{1}{2} + \tfrac{1}{2} > + \mid \tfrac{1}{2} - \tfrac{1}{2} >)$$

$$\begin{pmatrix} 0 \\ 1 \end{pmatrix} \equiv \tfrac{1}{\sqrt{2}}(\mid \tfrac{1}{2} + \tfrac{1}{2} > - \mid \tfrac{1}{2} - \tfrac{1}{2} >) \tag{4.19}$$

Then:

$$\left|\tfrac{1}{2}+\tfrac{1}{2}\right> = \tfrac{1}{\sqrt{2}}\begin{pmatrix}1\\1\end{pmatrix} \quad \text{and} \quad \left|\tfrac{1}{2}-\tfrac{1}{2}\right> = \tfrac{1}{\sqrt{2}}\begin{pmatrix}1\\-1\end{pmatrix} \tag{4.20}$$

It is not to difficult to spot the new representation of the generator J_3. Alternatively we can find the new representation by recalling that the matrix of eigenvectors E for a matrix M relates to the diagonal matrix of eigenvalues Λ by $ME = E\Lambda$. For a Hermitian matrix M, Λ is real and E is unitary.[a] Hence $M = E\Lambda E^H$. Therefore, associating Λ with the eigenvalues and E with the vectors in Eq. (4.20):

$$J_3 = \tfrac{1}{\sqrt{2}}\begin{pmatrix}1&1\\1&-1\end{pmatrix}\begin{pmatrix}\tfrac{1}{2}&0\\0&-\tfrac{1}{2}\end{pmatrix}\tfrac{1}{\sqrt{2}}\begin{pmatrix}1&1\\1&-1\end{pmatrix} = \begin{pmatrix}0&\tfrac{1}{2}\\\tfrac{1}{2}&0\end{pmatrix} \tag{4.21}$$

To find J_+ and J_-, recall:

$$J_+\left|\tfrac{1}{2}+\tfrac{1}{2}\right> = 0$$
$$J_+\left|\tfrac{1}{2}-\tfrac{1}{2}\right> = \left|\tfrac{1}{2}+\tfrac{1}{2}\right> \tag{4.22}$$
$$J_- = J_+{}^H$$

So by inspection:

$$J_+ = \tfrac{1}{2}\begin{pmatrix}1&-1\\1&-1\end{pmatrix} \quad \text{and} \quad J_- = \tfrac{1}{2}\begin{pmatrix}1&1\\-1&-1\end{pmatrix} \tag{4.23}$$

Hence:

$$J_1 = \begin{pmatrix}\tfrac{1}{2}&0\\0&-\tfrac{1}{2}\end{pmatrix} \quad \text{and} \quad J_2 = \begin{pmatrix}0&\tfrac{i}{2}\\\tfrac{-i}{2}&0\end{pmatrix} \tag{4.24}$$

So with this basis:

$$J_1 = \tfrac{1}{2}\sigma_3, \quad J_2 = \tfrac{-1}{2}\sigma_2, \quad J_3 = \tfrac{1}{2}\sigma_1 \tag{4.25}$$

---

[a] By considering $(ME)^H E$ and $E^H(ME)$, you can show that Λ is real and E is unitary. You will need to remember that Λ is diagonal by construction and look at components.

It looks like a "rotation" about the "1 = 3 axis". With a different choice of basis, we would not necessarily obtain the Pauli matrices.

4.3 A 3D Representation

A 3D representation corresponds to operators in a space with 3 base vectors. A suitable basis can be found from the states corresponding to $j = 1$ with $m = -1, 0, +1$. We can choose base vectors:

$$\begin{pmatrix} 1 \\ 0 \\ 0 \end{pmatrix} \equiv |1+1>, \quad \begin{pmatrix} 0 \\ 1 \\ 0 \end{pmatrix} \equiv |10>, \quad \begin{pmatrix} 0 \\ 0 \\ 1 \end{pmatrix} \equiv |1-1> \qquad (4.26)$$

Then by observation:

$$J_3 = \begin{pmatrix} 1 & 0 & 0 \\ 0 & 0 & 0 \\ 0 & 0 & -1 \end{pmatrix} \qquad (4.27)$$

It is easy to construct J_+ and J_- by observation, and as before:

$$J_1 = \tfrac{1}{\sqrt{2}} \begin{pmatrix} 0 & 1 & 0 \\ 1 & 0 & 1 \\ 0 & 1 & 0 \end{pmatrix} \quad \text{and} \quad J_2 = \tfrac{i}{\sqrt{2}} \begin{pmatrix} 0 & -1 & 0 \\ 1 & 0 & -1 \\ 0 & 1 & 0 \end{pmatrix} \qquad (4.28)$$

It is easy to verify that these matrices satisfy the SU(2) Lie algebra of Eq. (4.5). This is a consequence of the construction of the states using the raising & lowering operators, whose properties in turn were derived from the SU(2) Lie algebra – see Probs 4.1 and 4.2.

So we see SU(n) can have representations in vector spaces of various dimensions – in each dimension, there will be an infinite number of possible representations. The lowest dimensional representation is in nD. The matrices which correspond to the simple base vectors in nD, namely $(1, 0, 0, \ldots)$, $(0, 1, 0, \ldots)$, $(0, 0, 1, \ldots)$ etc., are called the **<u>fundamental</u>**

representation, of which there are $(n-1)$ representations. For SU(2), there is just one fundamental representation, namely $\{\sigma_1/2,\ \sigma_2/2,\ \sigma_3/2\}$. For $n > 2$, there are more quantum numbers and additional ways in which they can be associated to base vectors.

4.4 Rotations

4.4.1 *Comparison between spin and spatial rotations*

Transformations under SU(2) are performed by operators of the form $\exp(i\theta_i J_i)$ (implicit sum over $i = 1, 3$), where θ_i can be written as θn_i, and where n_i are the components of a normal vector $\boldsymbol{n}$. In 2D, we have identified the generators J_i with the Pauli spin matrices $\sigma_i/2$ which correspond to the spin-$^1/_2$ angular momentum operators. Furthermore, the operators have the form we would expect from our consideration of 3D transformations of spatial wavefunctions in QM (see Chapter 1) – i.e. the properties of the SU(2) generators J_i are similar to those of the angular momentum operators L_i, which are the generators of spatial rotations. By this, we mean that the Lie algebra of the operators is the same, as are the eigenstates and eigenvalues. The features of spatial and spin rotations are summarised in Table 4.1. They correspond to the groups SO(3) and SU(2) respectively. These two groups are isomorphic.

Table 4.1 Comparison of spatial and spin rotations.

	Spatial Rotations	**Spin Rotations**
Vector space	$\psi(x)$ – spatial w/f	ψ – spinor
Operator and appropriate representation	$\exp(i\theta\boldsymbol{n}\cdot\boldsymbol{L})$ where $\boldsymbol{L} = \boldsymbol{x}\times(-i\nabla)$; operator is a scalar	$\exp(i\theta\boldsymbol{n}\cdot\boldsymbol{J})$ where $\boldsymbol{J}$ is a set of 3 matrices; operator is a matrix
Operates on	x – 3D space	spin vector
Symmetry group	SO(3)	SU(2)

4.4.2 *Combination of spin rotations*

The transformation operator for spin-$\frac{1}{2}$ states can be written as $U = \exp(i\theta_i\sigma_i/2)$. Writing $\theta_i = 2\omega n_i$ gives $U = \exp(i\omega\boldsymbol{n}\cdot\boldsymbol{\sigma})$, where $\boldsymbol{\sigma}$ is understood to be the "vector" of the three Pauli matrices, and the scalar product is indicated explicitly by "·". Expanding the exponential:

$$U = 1 + (i\omega\boldsymbol{n}\cdot\boldsymbol{\sigma}) + (i\omega\boldsymbol{n}\cdot\boldsymbol{\sigma})^2/2! + (i\omega\boldsymbol{n}\cdot\boldsymbol{\sigma})^3/3! + \ldots \qquad (4.29)$$

Now:

$$(\boldsymbol{n}\cdot\boldsymbol{\sigma})^2 = n_1^2\sigma_1^2 + \ldots + n_1 n_2(\sigma_1\sigma_2 + \sigma_2\sigma_1) + \ldots$$

Using:

$$\sigma_1^2 = 1 \quad \text{and} \quad \sigma_1\sigma_2 + \sigma_2\sigma_1 = 0$$

gives:

$$(\boldsymbol{n}\cdot\boldsymbol{\sigma})^2 = n_1^2 + n_2^2 + n_3^2 = 1 \qquad (4.30)$$

So:

$$U = \{1 - \omega^2/2! + \omega^4/4! - \ldots\} + i\{\omega - \omega^3/3! + \omega^5/5! - \ldots\}\boldsymbol{n}\cdot\boldsymbol{\sigma}$$

$$= \cos\omega + i\sin\omega\,\boldsymbol{n}\cdot\boldsymbol{\sigma} \qquad (4.31)$$

So:

$$\exp(i\tfrac{\theta}{2}\boldsymbol{n}\cdot\boldsymbol{\sigma}) = \cos\tfrac{\theta}{2} + i\sin\tfrac{\theta}{2}\,\boldsymbol{n}\cdot\boldsymbol{\sigma} \qquad (4.32)$$

where the cosine term multiplies an implicit unit matrix.

We can now prove the Lie nature of SU(2) explicitly. We should expect that the product of two operators is a third one of a similar form, where the parameters of the result are a function of the parameters of the first two operations. Consider the product of two operations with parameters α and β and normal vectors $\boldsymbol{a}$ and $\boldsymbol{b}$:

$$\exp(i\alpha\boldsymbol{a}\cdot\boldsymbol{\sigma})\exp(i\beta\boldsymbol{b}\cdot\boldsymbol{\sigma})$$
$$= (\cos\alpha + i\sin\alpha\,\boldsymbol{a}\cdot\boldsymbol{\sigma})(\cos\beta + i\sin\beta\,\boldsymbol{b}\cdot\boldsymbol{\sigma}) \qquad (4.33)$$

Using Eq. (4.4), $a \cdot \sigma \, b \cdot \sigma = a \cdot b + i a \times b \cdot \sigma$, so the product is:

$$\cos \alpha \cos \beta - \sin \alpha \sin \beta \, (a \cdot b + i a \times b \cdot \sigma) +$$
$$i(\sin \alpha \cos \beta \, a \cdot \sigma + \sin \beta \cos \alpha \, b \cdot \sigma)$$
$$= \{\cos \alpha \cos \beta - \sin \alpha \sin \beta \, a \cdot b\} +$$
$$i\{a \times b + \sin \alpha \cos \beta \, a + \sin \beta \cos \alpha \, b\} \cdot \sigma \qquad (4.34)$$

We can write the last expression as $C + i \, S \cdot \sigma$, where S is 3D vector. To show that this has the form $\cos \gamma + i \sin \gamma \, c \cdot \sigma$, it is sufficient to show that $C^2 + S^2 = 1$ – this is straightforward, albeit tedious, and is left as an exercise for the enthusiastic reader. So:

$$\cos \gamma = \cos \alpha \cos \beta - \sin \alpha \sin \beta \, a \cdot b \quad \text{and} \quad c = S / \sin \gamma \qquad (4.35)$$

Hence we have found the 3 parameters of the third operation (γ, c) in terms of the parameters of the first two operations (α, a) and (β, b), as required for a Lie group.

4.4.3 *Rotation matrices*

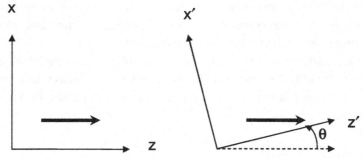

Fig. 4.1 Rotation of axes about the *y*-axis by an angle θ.

Consider an object, for example a vector representing the spin of a particle $|j\,m>$, embedded in a reference frame, as shown in Fig. 4.1. As before, j and m are the eigenvalues of J^2 and J_3. Now if we undertake a rotation by an angle θ about the *y*-axis, then the transformation to the new frame is given by:

$$|jm> \rightarrow R_y(\theta) | jm > \qquad (4.36)$$

where:

$$R_y(\theta) = \exp(i\theta J_y) \qquad (4.37)$$

and J_y (equivalent to J_2) is the second component of the angular momentum operator (could be spin or orbital, although we will be thinking about spin). Note: because J^2 commutes with J_y and hence R_y (a power series of J_y), the j quantum number is unchanged. However, because $J_y \sim (J_+ - J_-)$, a state which initially has a well-defined value of J_3 will be mapped on to an admixture of states with different m values.

The **rotation matrices** $d(\theta)$ are defined by:

$$d^j_{mm'}(\theta) = < jm' \mid R_y(\theta) \mid jm > \qquad (4.38)$$

Note the order $m \rightarrow m'$. The components of these matrices can be found in the *Particle Data Group (PDG) Book* [PDG]. The significance of these matrices is that the modulus squared is the probability that a particle with $J_3 = m$ will have $J_3 = m'$ after the rotation to the new frame.

Why is the rotation done about the y-axis? It leads to real matrix elements. If we were to use the x-axis, we would get the same physical probabilities once the amplitudes were squared.

These matrices are nice since by quick reference to the *PDG Book*, insight can be obtained into angular distributions for decays or scattering processes without having to calculate the amplitudes from the Feynman diagrams.

4.4.3.1 *Spin 0*

Consider the decay of a spin-0 object, such as a π^0, decaying to two photons. The pion has no spin vector and has no preferred orientation in space. Therefore we should expect the photons to emerge back-to-back but in any direction with respect to any arbitrary axes. We would expect the angular distribution to be flat in phase-space as a function of $\cos\theta$. If we think about the rate for the photons to emerge along the z-axis, and then consider a different frame, rotated by an angle θ, as illustrated in

Fig. 4.2, then the amplitudes should be related by the rotation matrix found in the *PDG Book* [PDG]:

$$d_{00}^0(\theta) = 1 \tag{4.39}$$

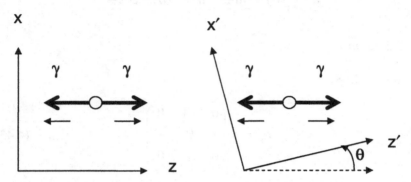

Fig. 4.2 π^0 decaying to two photons in different frames.

So, as expected, we see that the decay amplitudes at any angle (more precisely, as a function of $\cos\theta$) are the same.

4.4.3.2 Spin $^1/_2$

From Eq. (4.32), for spin $^1/_2$:

$$R_y(\theta) = \cos\tfrac{\theta}{2} + i\sin\tfrac{\theta}{2}\sigma_2 = \cos\tfrac{\theta}{2}\begin{pmatrix} 1 & 0 \\ 0 & 1 \end{pmatrix} + i\sin\tfrac{\theta}{2}\begin{pmatrix} 0 & -i \\ i & 0 \end{pmatrix}$$
$$= \begin{pmatrix} \cos\tfrac{\theta}{2} & \sin\tfrac{\theta}{2} \\ -\sin\tfrac{\theta}{2} & \cos\tfrac{\theta}{2} \end{pmatrix} \tag{4.40}$$

So:

$$|\tfrac{1}{2} + \tfrac{1}{2} > \equiv \begin{pmatrix} 1 \\ 0 \end{pmatrix} \rightarrow \cos\tfrac{\theta}{2}\begin{pmatrix} 1 \\ 0 \end{pmatrix} - \sin\tfrac{\theta}{2}\begin{pmatrix} 0 \\ 1 \end{pmatrix}$$
$$|\tfrac{1}{2} - \tfrac{1}{2} > \equiv \begin{pmatrix} 0 \\ 1 \end{pmatrix} \rightarrow \sin\tfrac{\theta}{2}\begin{pmatrix} 1 \\ 0 \end{pmatrix} + \cos\tfrac{\theta}{2}\begin{pmatrix} 0 \\ 1 \end{pmatrix} \tag{4.41}$$

Hence:

$$d^{\frac{1}{2}}_{+\frac{1}{2}+\frac{1}{2}}(\theta) = \cos\frac{\theta}{2} \quad d^{\frac{1}{2}}_{+\frac{1}{2}-\frac{1}{2}}(\theta) = -\sin\frac{\theta}{2}$$

$$d^{\frac{1}{2}}_{-\frac{1}{2}+\frac{1}{2}}(\theta) = \sin\frac{\theta}{2} \quad d^{\frac{1}{2}}_{-\frac{1}{2}-\frac{1}{2}}(\theta) = \cos\frac{\theta}{2} \tag{4.42}$$

4.4.3.3 *Spin 1*

From Eq. (4.28):

$$J_y = \frac{i}{\sqrt{2}} \begin{pmatrix} 0 & -1 & 0 \\ 1 & 0 & -1 \\ 0 & 1 & 0 \end{pmatrix} \tag{4.43}$$

The rotation matrix can be found from the power series for the exponential. It is readily shown that:

$$J_y^{\,2} = J_y^{\,4} = J_y^{\,6}... \quad \text{and} \quad J_y = J_y^{\,3} = J_y^{\,5}... \tag{4.44}$$

From this it is easy to show that:

$$R_y(\theta) = \frac{1}{2}\begin{pmatrix} 1+\cos\theta & 0 & 1-\cos\theta \\ 0 & 2\cos\theta & 0 \\ 1-\cos\theta & 0 & 1+\cos\theta \end{pmatrix} +$$

$$\frac{1}{\sqrt{2}}\begin{pmatrix} 0 & \sin\theta & 0 \\ -\sin\theta & 0 & \sin\theta \\ 0 & -\sin\theta & 0 \end{pmatrix} \tag{4.45}$$

Hence:

$$d^{1}_{+1+1}(\theta) = \tfrac{1}{2}(1+\cos\theta) \quad d^{1}_{+10}(\theta) = \frac{-1}{\sqrt{2}}\sin\theta$$

$$d^{1}_{+1-1}(\theta) = \tfrac{1}{2}(1-\cos\theta) \quad d^{1}_{00}(\theta) = \cos\theta \qquad \text{etc.} \tag{4.46}$$

As an example, we can consider the production of W-bosons in proton-antiproton colliders: $u\bar{d} \to W^+ \to e^+v$ as illustrated in Fig. 4.3.

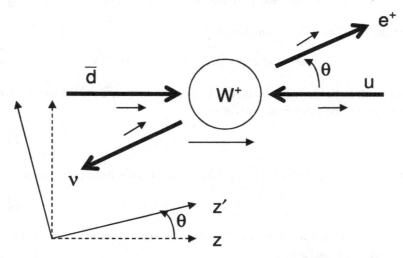

Fig. 4.3 Production and decay of W-boson. Spin directions are shown by the smaller arrows.

The initial state consists of two aligned spin-$^1/_2$ states in a ($j=1$, $m=1$) configuration;[b] so does the final state, albeit with respect to a different axis. What is the projection of the initial |11> state in the initial frame onto a |11> state in the final frame? It is simply the rotation matrix $d^1_{+1+1}(\theta)$. So the angular distribution of the decay is:

$$\frac{dN}{d\cos\theta} \sim (d^1_{+1\ +1}(\theta))^2 \sim (1+\cos\theta)^2 \tag{4.47}$$

This is the famous V-A distribution.

4.5   Gauge Transformations and the Adjoint Representation

Let us consider gauge transformations in **non-Abelian Groups**. We consider local transformations of the form $exp(i\alpha_i(x)X_i)$, where X_i are the generators and there is an implicit summation over the index i.

---

[b] To interact via the weak interaction, fermions must be left-handed and anti-fermions, right-handed.

Under an infinitesimal gauge transformation:

$$\psi \rightarrow \exp(i\alpha_i X_i)\psi = (1 + i\alpha_i X_i)\psi \qquad (4.48)$$

and the conjugate state transforms:

$$\overline{\psi} \rightarrow \overline{\psi}\exp(-i\alpha_i X_i) = \overline{\psi}(1 - i\alpha_i X_i) \qquad (4.49)$$

To maintain the invariance of the Lagrangian $L \sim \overline{\psi}\partial\!\!\!/\psi$, the derivative is replaced by the covariant derivative:

$$D^\mu = \partial^\mu - iX_i W_i^\mu \qquad (4.50)$$

where a vector field W^μ has been introduced. This field transforms:

$$W^\mu \rightarrow W^\mu + \delta W^\mu \qquad (4.51)$$

The Lagrangian (dropping the "slash" corresponding to contraction with γ matrices) transforms:

$$L \rightarrow \overline{\psi}(1 - i\alpha_i X_i)(\partial - iX_j W_j - iX_j \delta W_j)(1 + i\alpha_k X_k)\psi \qquad (4.52)$$

where distinct vector indices are introduced and the Lorentz indices are dropped. Expanding to first order in α and δW (remember that the parameters α and the fields W commute; the only non-commuting items are the generators):

$$L \rightarrow \overline{\psi}(\partial - iX_j W_j - iX_j \delta W_j + i(\partial\alpha_k)X_k - \alpha_i X_i X_j W_j +$$
$$\alpha_k X_j W_j X_k)\psi \qquad (4.53)$$
$$= \overline{\psi}(\partial - iX_j W_j - iX_j \delta W_j + i(\partial\alpha_k)X_k - \alpha_i[X_i, X_j]W_j)\psi$$

If this is to be invariant:

$$-iX_j \delta W_j + i(\partial\alpha_k)X_k - \alpha_i[X_i, X_j]W_j = 0 \qquad (4.54)$$

The indices are dummy, so we can replace them. Also, we can replace $[X_i, X_j]$ with $if_{ijk}X_k$ – where the f's are the structure constants:

$$iX_k \delta W_k = i(\partial\alpha_k)X_k - i\alpha_i f_{ijk} X_k W_j \qquad (4.55)$$

Equating all the coefficients of the X's (they form a basis):

$$\delta W_k = \partial \alpha_k - f_{ijk} \alpha_i W_j = \partial \alpha_k - iT^i_{\ jk} \alpha_i W_j \qquad (4.56)$$

where T is the adjoint (see Section 2.3.4). Therefore we can say that the gauge field "transforms according to the adjoint representation".[c]

In SU(2), the structure constants are the elements of the Levi–Civita tensor – see Eq. (4.5). Since the Levi–Civita tensor corresponds to the generators for SO(3) (rotations in 3D) (see Prob. 2.5), the 3 gauge fields, W, transform like a **vector** under the gauge transformations. Since the fields form a multiplet of multiplicity of 3 (they transform amongst themselves as indicated in Eq. (4.56)), we conclude they form an $I = 1$ triplet (from this point, I will use I rather than J, since we are explicitly referring to isospin). The $I = 1$ designation can be confirmed explicitly by considering the square of the generator – see the comments on Prob. 2.6 in the Appendix.

For particular quantum numbers, we consider a suitable vector space and associate the matter fields (fermion, spin $^1/_2$) with the base vectors of the vector space – sometimes called confusingly the **fundamental representation**. The associated intermediate gauge bosons are given in Table 4.2 for several symmetries. The gauge bosons are directly related to the generators, so as we saw in Section 2.5.1, there will be n^2-1 bosons for SU(n).

Table 4.2 Symmetries and their gauge bosons for particular base vectors.

Local Symmetry	Base Vectors	Gauge Bosons
Nuclear Isospin – SU(2)	(n, p)	π^+, π^0, π^-
Quark Isospin – SU(2)	(u, d)	π^+, π^0, π^-
Weak Isospin – SU(2)	$(u, d), (v, e)$ etc.	W^+, W^0, W^-
QCD – SU(3)	(r, b, g)	8 gluons

---

[c] We ignore the $\partial \alpha$ term, which is complicated for non-Abelian groups.

4.6 Isospin

4.6.1 *Hadronic isospin*

Experimentally, the Hamiltonian describing nucleon interactions is (fairly) independent of whether a nucleon is a proton or a neutron – it has a **symmetry** which is called <u>isospin</u>. As far as strong interactions are concerned, the proton and neutron are indistinguishable and hence in QM, they are interchangeable. This interchange is brought about by the gauge bosons of the symmetry, as illustrated in Fig. 4.4. These bosons are the familiar pions.

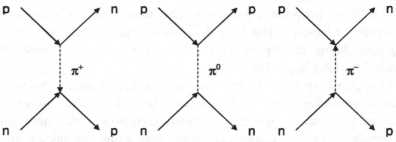

Fig. 4.4 Pion exchange between nucleons.

As we saw in Section 4.5, the gauge bosons (pions) form an $I = 1$ triplet under SU(2). Therefore the symmetry between the 3 pions within SU(2), means that in high-energy hadron collisions, the pions are produced in approximately equal numbers.

We can treat the nucleon as a particle which has two possible states: p or n, and we denote the state by a vector (p, n). In terms of quantum numbers (I, I_3) – along the same lines as the angular momentum labels – we have:

Doublet $p = (^1/_2, +^1/_2)$ $n = (^1/_2, -^1/_2)$

Triplet $\pi^+ = (1, +1)$ $\pi^0 = (1, 0)$ $\pi^- = (1, -1)$

In reality, the hadrons are distinguishable (else we would not have been able to label them) – they are distinguished by their electric charge and mass. These break the symmetry of the Hamiltonian, but since the electromagnetic effects are ~1/100th the size of the strong interaction, the symmetry is fairly good.

By observation, we can write the charge operator:

$$Q = \tfrac{1}{2}B + I_3 \qquad (4.57)$$

Any complete Hamiltonian contains a charge operator Q. Since Q commutes with I_3 and I^2, so I_3 and I^2 will correspond to conserved quantum numbers. However Q does not commute with I (vector), and so will not be invariant under general isospin transformations, such as those corresponding to p–n interchange – this requires a raising or lowering operator $I_\pm = I_1 \pm iI_2$.

Historically, a lot was made of this symmetry to understand nucleon scattering and production. We will examine how one can use the "machinery" when we look at the quark model. As an example, the **deuteron** is made from a proton and a neutron in a ground-state s-wave (symmetric spatial wave-function). It appears as a single state (there are no other comparable particles) and thus is an $I = 0$ antisymmetric state. To ensure the complete wave-function for two "identical" particles is antisymmetric, the spins must be in a symmetric state, namely $J = 1$.

4.6.2 *Quark isospin*

The reason that isospin is a good symmetry at the nucleon level is because of the underlying symmetry associated with the quarks, in particular $\{u, d\}$:

- The principle interactions are via QCD and these are flavour-blind.
- They have similar (negligible) masses.

Consequently there is an SU(2) isospin symmetry associated with the base states $\{u, d\}$. The adjoint can be associated with exchange pions, as shown in Fig. 4.5 – although this is actually a "trivial" exchange of quarks which is found predominantly at lower energies, while at higher energies, the principle exchange is via gluons – the gauge bosons of SU(3)$_{colour}$, acting between colour charges.

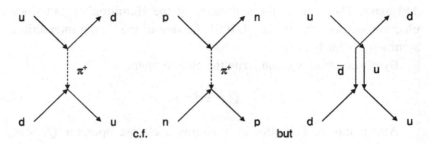

Fig. 4.5 Pion exchange between quarks. The first diagram shows the pion exchange at the quark level; the second shows what happens at the nucleon level (see Fig. 4.4); the third reveals what is actually happening at the quark level when the pion is understood in terms of quarks.

SU(2) for strong interactions of quarks is less useful. However, it will prove useful for hadron classification.

4.6.3 *Weak isospin*

Weak isospin is an exact symmetry of the SM (by construction). The base vectors are formed from ($I_3 = +^1/_2$, $I_3 = -^1/_2$) doublets – there are 6 particle states which have these quantum numbers:

$$\begin{pmatrix} u \\ d \end{pmatrix}_L, \ \begin{pmatrix} c \\ s \end{pmatrix}_L, \ \begin{pmatrix} t \\ b \end{pmatrix}_L, \ \begin{pmatrix} v_e \\ e^- \end{pmatrix}_L, \ \begin{pmatrix} v_\mu \\ \mu^- \end{pmatrix}_L, \ \begin{pmatrix} v_\tau \\ \tau^- \end{pmatrix}_L \qquad (4.58)$$

Note, all these are left-handed fermion states: left-handed refers to the **chirality**, corresponding to the projection operator $\frac{1}{2}(1-\gamma_5)$. The local SU(2) gauge symmetry leads to a vector of fields: $\{W_1, W_2, W_3\}$. These relate to the fields $\{W^+, W^0, W^-\}$, where:

$$W^\pm = \frac{1}{\sqrt{2}}(W_1 \mp iW_2) \qquad (4.59)$$

and the W^0 field is transformed into the Z-boson field (see Chapter 8).

4.7 Conjugate States

In SU(2), we define the **base state** (sometimes called the "fundamental representation"):

$$2 \equiv \begin{pmatrix} u \\ d \end{pmatrix} \tag{4.60}$$

We would like to form a **conjugate state** which will correspond to the antiquarks and which will have *convenient* transformation properties. The conjugate will contain the antiquark wave-functions, and to be consistent with normal properties, we expect the quantum numbers of the antiquarks to be negated with respect to those of the quarks. This is also consistent with the expression $Q = \frac{1}{2} B + I_3$ (in the case of quark isospin). The quantum numbers are shown in Table 4.3.

Table 4.3 Quantum numbers for u and d quarks.

	Q	B	$I_3 = Q - \frac{1}{2} B$
u	$+\frac{2}{3}$	$+\frac{1}{3}$	$+\frac{1}{2}$
$\bar{u}$	$-\frac{2}{3}$	$-\frac{1}{3}$	$-\frac{1}{2}$
d	$-\frac{1}{3}$	$+\frac{1}{3}$	$-\frac{1}{2}$
$\bar{d}$	$+\frac{1}{3}$	$-\frac{1}{3}$	$+\frac{1}{2}$

Antiparticles behave like the complex conjugates:[d]

$$\bar{u} \sim (u)^* \text{ and } \bar{d} \sim (d)^* \tag{4.61}$$

Under SU(2), the **2** transforms like:

$$\begin{pmatrix} u' \\ d' \end{pmatrix} = \exp(i\tfrac{\theta}{2} \boldsymbol{n} \cdot \boldsymbol{\sigma}) \begin{pmatrix} u \\ d \end{pmatrix} = (\cos\tfrac{\theta}{2} + i\sin\tfrac{\theta}{2} \boldsymbol{n} \cdot \boldsymbol{\sigma}) \begin{pmatrix} u \\ d \end{pmatrix} \tag{4.62}$$

---

[d] Creation and annihilation operators in quantum field theory are the complex conjugate of each other [Halzen & Martin, Section 5.4].

Taking the complex conjugate (not Hermitian):[e]

$$\begin{pmatrix} u'\,* \\ d'\,* \end{pmatrix} = (\cos\tfrac{\theta}{2} - i\sin\tfrac{\theta}{2}\boldsymbol{n}\cdot\boldsymbol{\sigma}^*)\begin{pmatrix} u\,* \\ d\,* \end{pmatrix} \tag{4.63}$$

So the transformation of the antiquark states is given by:

$$\begin{pmatrix} \bar{u}' \\ \bar{d}' \end{pmatrix} = (\cos\tfrac{\theta}{2} - i\sin\tfrac{\theta}{2}\boldsymbol{n}\cdot\boldsymbol{\sigma}^*)\begin{pmatrix} \bar{u} \\ \bar{d} \end{pmatrix} \tag{4.64}$$

It will prove to be convenient if the conjugate state will have the same transformation properties in SU(2) as the **2**. To this end, we define the conjugate to be:

$$\bar{\mathbf{2}} = M\begin{pmatrix} \bar{u} \\ \bar{d} \end{pmatrix} \tag{4.65}$$

where M is a 2×2 matrix and we would like:

$$M\begin{pmatrix} \bar{u}' \\ \bar{d}' \end{pmatrix} = (\cos\tfrac{\theta}{2} + i\sin\tfrac{\theta}{2}\boldsymbol{n}\cdot\boldsymbol{\sigma})M\begin{pmatrix} \bar{u} \\ \bar{d} \end{pmatrix} \tag{4.66}$$

The challenge is to find a suitable M. Pre-multiplying Eq. (4.64) by M:

$$M\begin{pmatrix} \bar{u}' \\ \bar{d}' \end{pmatrix} = M(\cos\tfrac{\theta}{2} - i\sin\tfrac{\theta}{2}\boldsymbol{n}\cdot\boldsymbol{\sigma}^*)\begin{pmatrix} \bar{u} \\ \bar{d} \end{pmatrix} \tag{4.67}$$

So:

$$M(\cos\tfrac{\theta}{2} - i\sin\tfrac{\theta}{2}\boldsymbol{n}\cdot\boldsymbol{\sigma}^*) = (\cos\tfrac{\theta}{2} + i\sin\tfrac{\theta}{2}\boldsymbol{n}\cdot\boldsymbol{\sigma})M$$
$$\Rightarrow -M\boldsymbol{\sigma}^* = \boldsymbol{\sigma}M \Rightarrow \boldsymbol{\sigma}^* = -M^{-1}\boldsymbol{\sigma}M \tag{4.68}$$

---

[e] We take the complex conjugate, not the Hermitian conjugate, since we want to understand how the column vector, not the row vector, transforms.

Recalling that σ is actually a vector of the three Pauli matrices:

$$\sigma^* = \begin{pmatrix} \sigma_1 \\ -\sigma_2 \\ \sigma_3 \end{pmatrix} = \begin{pmatrix} -M^{-1}\sigma_1 M \\ -M^{-1}\sigma_2 M \\ -M^{-1}\sigma_3 M \end{pmatrix} \qquad (4.69)$$

Bearing in mind that $\sigma_i \sigma_j = \sigma_j \sigma_i$ for $i = j$ and $\sigma_i \sigma_j = -\sigma_j \sigma_i$ for $i \neq j$, we can satisfy the above if $M = phase \times \sigma_2$ (consider the terms $\sigma_i M$ and consequences of commuting the two terms to remove M via $M^{-1}M = I$). We take:

$$M = \begin{pmatrix} 0 & +1 \\ -1 & 0 \end{pmatrix} \qquad (4.70)$$

so:

$$\overline{\mathbf{2}} = \begin{pmatrix} \overline{d} \\ -\overline{u} \end{pmatrix} \qquad (4.71)$$

will transform like:

$$\mathbf{2} \equiv \begin{pmatrix} u \\ d \end{pmatrix} \qquad (4.72)$$

We see that $\overline{d}$ is at the top of the $\overline{\mathbf{2}}$, consistent with having $I_3 = {}^1/_2$ as indicated in Table 4.3.

This result will turn out to be very useful for the formation of $q\overline{q}$ states. It arises from some good fortune in SU(2) and is not the case in other SU(n) groups.

4.8 Problems

Prob. 4.1* Show from the properties of J_3 and the raising & lowering operators $J_{\pm}$, as defined in Eqs (4.10) and (4.11) & (4.12), respectively, that one can recover the Lie algebra expressed in Eq. (4.5).

Prob. 4.2 Verify that the matrices of Eqs (4.27) & (4.28) do satisfy the SU(2) Lie algebra.

Prob. 4.3 In some GUT Models, the proton decays $p \rightarrow \pi^0 e^+$. Suppose a proton is polarised with its spin upwards, then the angular distribution of the positrons will depend on the amplitudes for the various helicity states. Calculate the angular distribution of the right-handed positrons.

Chapter 5

Combining Fermions

5.1 Introduction

In this chapter, we will see how to form **combined states**, in particular in the context of **hadron wavefunctions**. The focus will be primarily on SU(2).

The key message to extract from this chapter is:

- Multiplets of states, associated with irreducible representations, correspond to sets of observed particle states with similar properties.

5.2 Combining States

We wish to construct multi-particle states which have well-defined properties under group transformations. The first question is: how do we undertake transformations on multi-particles states? The transformations we wish to consider are on internal degrees of freedom (associated with quantum numbers, rather than space-time). Nevertheless, it is helpful to think of the analogy with spatial transformations, for example rotations. Consider a state corresponding to two independent particles a and b:

$$\psi^{ab}(x) = \psi^a(x)\psi^b(x) \tag{5.1}$$

The wavefunctions are functions of the common space-time. If we consider spatial rotations by an angle θ, about a particular axis, corresponding to a unit vector n, with the corresponding angular momentum operator $L = n \cdot x \times (-i\nabla)$, then the transformed state is:

$$(\psi^{ab})'(x) = \exp(i\theta L)\psi^a(x)\psi^b(x) = \sum_p \frac{(i\theta L)^p}{p!}\psi^a(x)\psi^b(x) \tag{5.2}$$

From the product rule for differentiation:

$$L\psi^a(x)\psi^b(x) = \{L\psi^a(x)\}\psi^b(x) + \psi^a(x)\{L\psi^b(x)\}$$
$$= (L^a + L^b)\psi^a(x)\psi^b(x) \tag{5.3}$$

$$L^2\psi^a(x)\psi^b(x)$$
$$= \{L^2\psi^a(x)\}\psi^b(x) + 2\{L\psi^a(x)\}\{L\psi^b(x)\} + \psi^a(x)\{L^2\psi^b(x)\} \tag{5.4}$$
$$= (L^a + L^b)^2\psi^a(x)\psi^b(x)$$

where L^a is the angular momentum operator which is restricted to operate only on $\psi^a(x)$. Therefore:

$$(\psi^{ab})'(x) = \sum_p \frac{(i\theta(L^a + L^b))^p}{p!}\psi^a(x)\psi^b(x) = \exp(i\theta(L^a + L^b))\psi^{ab}(x) \tag{5.5}$$

So we find that the angular momentum operator for the combined state is equal to the sum of operators for individual states. We will assume the same for the generators in SU(2) and assume $\boldsymbol{J}^{ab} = \boldsymbol{J}^a + \boldsymbol{J}^b$ (vector relationship) and for the third component: $J_3^{ab} = J_3^a + J_3^b$. Therefore:

$$J^2 = J^{a^2} + J^{b^2} + 2\boldsymbol{J}^a \cdot \boldsymbol{J}^b$$
$$= J^{a^2} + J^{b^2} + 2(J_1^a J_1^b + J_2^a J_2^b + J_3^a J_3^b) \tag{5.6}$$
$$= J^{a^2} + J^{b^2} + J_+^a J_-^b + J_-^a J_+^b + 2J_3^a J_3^b$$

and:

$$J_\pm = J_\pm^a + J_\pm^b \tag{5.7}$$

We label the states:

$$u \equiv |\tfrac{1}{2} + \tfrac{1}{2}> \quad \text{and} \quad d \equiv |\tfrac{1}{2} - \tfrac{1}{2}> \tag{5.8}$$

– think of these as spin or isospin (nucleon or quark flavour) states. To create the states formed from combinations, we start with the fully aligned state:

$$\psi^{ab} = \psi^a\psi^b \quad \text{with} \quad \psi^a = u \quad \text{and} \quad \psi^b = u \tag{5.9}$$

So:

$$J_3 uu = (J_3{}^a + J_3{}^b)uu = \{J_3{}^a u\}u + u\{J_3{}^b u\}$$
$$= \{\tfrac{1}{2}u\}u + u\{\tfrac{1}{2}u\} = uu \qquad (5.10)$$

and:

$$J^2 uu = (J^{a^2} + J^{b^2} + J_+{}^a J_-{}^b + J_-{}^a J_+{}^b + 2J_3{}^a J_3{}^b)uu$$
$$= (\tfrac{3}{4} + \tfrac{3}{4} + 0 + 0 + 2 \cdot \tfrac{1}{2} \cdot \tfrac{1}{2})uu = 2uu = 1(1+1)uu \qquad (5.11)$$

We see this state has $(j, m) = (1, 1)$, and so:

$$|1 + 1 >= uu \qquad (5.12)$$

Other states can readily be obtained using J_-. With $J_- u = d$:

$$J_- uu = (J_-{}^a + J_-{}^b)uu = \{J_-{}^a u\}u + u\{J_-{}^b u\} = du + ud \qquad (5.13)$$

This state can be shown to have $(j, m) = (1, 0)$, and so:

$$|10 >= \tfrac{1}{\sqrt{2}}(ud + du) \qquad (5.14)$$

Finally, we can obtain:

$$|1 - 1 >= dd \qquad (5.15)$$

We expect 4 combinations, since 2 particles can each have 2 possible spin states. The missing combination which can be obtained by orthogonality is:

$$|00 > = \tfrac{1}{\sqrt{2}}(ud - du) \qquad (5.16)$$

5.2.1 *Multiplets*

The benefit of constructing multiplets consisting of states of a given j and well-defined m is that under SU(2) transformations, states transform to states within the same multiplet (irreducible representation) – and therefore the members of the multiplet have related properties, such as masses or decay properties.

A state $|\psi\rangle = |j\, m\rangle$ transforms to $|\psi'\rangle = U\,|\psi\rangle \equiv \exp(i\theta n \cdot J)\,|j\, m\rangle$ under SU(2). Since J^2 commutes with all the J_i generators and their powers, it will commute with $U \equiv \exp(i\theta n \cdot J)$, and thus the J^2 quantum number (j) of the transformed state will be unchanged.

If the (mass) Hamiltonian H is invariant under the SU(2) transformations, then $U^H H U = H$ or $[H, U] = 0$, and thus the mass of the transformed state (also in the multiplet) is:

$$M' = \langle\psi'|H|\psi'\rangle = \langle\psi U^H|H|U\psi\rangle$$
$$= \langle\psi|U^H H U|\psi\rangle = \langle\psi|H|\psi\rangle = M \qquad (5.17)$$

(Note: we do not add a dash to H, else $M' \equiv M$ because of the definition of H' irrespective of the invariance or otherwise of H. Instead, we are interested in the effect of H for new states.)

As an illustration, let us consider the effect of a rotation $\exp(i\boldsymbol{\omega}\cdot\boldsymbol{\sigma})$, where $\boldsymbol{\omega} = \omega\mathbf{n}$, on the SU(2) singlet for two fermions:

$$|00\rangle \sim ud - du = \begin{pmatrix}1\\0\end{pmatrix}\begin{pmatrix}0\\1\end{pmatrix} - \begin{pmatrix}0\\1\end{pmatrix}\begin{pmatrix}1\\0\end{pmatrix} \qquad (5.18)$$

(drop the $1/\sqrt{2}$ normalisation for simplicity). The rotation operator is:

$$\exp(i\boldsymbol{\omega}\cdot\boldsymbol{\sigma}) = \exp(i\boldsymbol{\omega}\cdot\boldsymbol{\sigma}^a)\exp(i\boldsymbol{\omega}\cdot\boldsymbol{\sigma}^b) \qquad (5.19)$$

where $\boldsymbol{\sigma}^a$ acts on the first particle etc.. So:

$$\exp(i\boldsymbol{\omega}\cdot\boldsymbol{\sigma})|00\rangle$$
$$= \exp(i\boldsymbol{\omega}\cdot\boldsymbol{\sigma}^a)\exp(i\boldsymbol{\omega}\cdot\boldsymbol{\sigma}^b)\{\begin{pmatrix}1\\0\end{pmatrix}_a\begin{pmatrix}0\\1\end{pmatrix}_b - \begin{pmatrix}0\\1\end{pmatrix}_a\begin{pmatrix}1\\0\end{pmatrix}_b\} \qquad (5.20)$$

Recalling that:

$$\exp(i\boldsymbol{\omega}\cdot\boldsymbol{\sigma}) = \cos\omega + i\sin\omega\,\mathbf{n}\cdot\boldsymbol{\sigma}$$
$$= \begin{pmatrix} \cos\omega + i\sin\omega\, n_3 & i\sin\omega\,(n_1 - in_2) \\ i\sin\omega\,(n_1 + in_2) & \cos\omega - i\sin\omega\, n_3 \end{pmatrix} \qquad (5.21)$$

then:

$$\exp(i\omega \cdot \boldsymbol{\sigma})|00> = \begin{pmatrix} \cos \omega + i \sin \omega\, n_3 & i \sin \omega\, (n_1 - in_2) \\ i \sin \omega\, (n_1 + in_2) & \cos \omega - i \sin \omega\, n_3 \end{pmatrix} -$$
$$\begin{pmatrix} i \sin \omega\, (n_1 - in_2) & \cos \omega + i \sin \omega\, n_3 \\ \cos \omega - i \sin \omega\, n_3 & i \sin \omega\, (n_1 + in_2) \end{pmatrix}$$

$$(5.22)$$

It is helpful to think what these vectors look like in terms of the base vectors. For example, if we look at the first vector for particle a:

$$\begin{pmatrix} \cos \omega + i \sin \omega\, n_3 \\ i \sin \omega\, (n_1 + in_2) \end{pmatrix} \equiv$$
$$(\cos \omega + i \sin \omega\, n_3)\begin{pmatrix} 1 \\ 0 \end{pmatrix} + i \sin \omega (n_1 + in_2)\begin{pmatrix} 0 \\ 1 \end{pmatrix}$$

$$(5.23)$$

So the coefficient of $\begin{pmatrix} 1 \\ 0 \end{pmatrix}\begin{pmatrix} 1 \\ 0 \end{pmatrix}$ in Eq. (5.22) is:

$$(\cos \omega + i \sin \omega\, n_3) \times i \sin \omega\, (n_1 - in_2) -$$
$$i \sin \omega\, (n_1 - in_2) \times (\cos \omega + i \sin \omega\, n_3) = 0$$

$$(5.24)$$

and the coefficient of $\begin{pmatrix} 1 \\ 0 \end{pmatrix}\begin{pmatrix} 0 \\ 1 \end{pmatrix}$ is:

$$(\cos \omega + i \sin \omega\, n_3)(\cos \omega - i \sin \omega\, n_3) -$$
$$i \sin \omega\, (n_1 - in_2) \times i \sin \omega\, (n_1 + in_2)$$
$$= (\cos^2 \omega + \sin^2 \omega\, n_3{}^2) + \sin^2 \omega\, (n_1{}^2 + n_2{}^2)$$
$$= 1$$

$$(5.25)$$

since n is a unit vector, and hence $n_1{}^2 + n_2{}^2 + n_3{}^3 = 1$. Similarly for the other coefficients. Hence:

$$\exp(i\omega \cdot \boldsymbol{\sigma})|00> = 0\begin{pmatrix} 1 \\ 0 \end{pmatrix}\begin{pmatrix} 1 \\ 0 \end{pmatrix} + 1\begin{pmatrix} 1 \\ 0 \end{pmatrix}\begin{pmatrix} 0 \\ 1 \end{pmatrix} - 1\begin{pmatrix} 0 \\ 1 \end{pmatrix}\begin{pmatrix} 1 \\ 0 \end{pmatrix} + 0\begin{pmatrix} 0 \\ 1 \end{pmatrix}\begin{pmatrix} 0 \\ 1 \end{pmatrix}$$
$$= \begin{pmatrix} 1 \\ 0 \end{pmatrix}\begin{pmatrix} 0 \\ 1 \end{pmatrix} - \begin{pmatrix} 0 \\ 1 \end{pmatrix}\begin{pmatrix} 1 \\ 0 \end{pmatrix} = |00>$$

$$(5.26)$$

So we see the |00> singlet state is unchanged by an SU(2) transformation.

Similarly, by considering the change of sign, it is easy to show the state:

$$|10> \sim ud + du = \begin{pmatrix}1\\0\end{pmatrix}\begin{pmatrix}0\\1\end{pmatrix} + \begin{pmatrix}0\\1\end{pmatrix}\begin{pmatrix}1\\0\end{pmatrix} \qquad (5.27)$$

transforms to an admixture of $j = 1$ states, as do the states:

$$|1+1> \sim uu = \begin{pmatrix}1\\0\end{pmatrix}\begin{pmatrix}1\\0\end{pmatrix} \quad \text{and} \quad |1-1> \sim dd = \begin{pmatrix}0\\1\end{pmatrix}\begin{pmatrix}0\\1\end{pmatrix} \qquad (5.28)$$

In general, a transformation $\exp(i\theta n \cdot J)$ can be expanded into a power series containing products of J_1, J_2, and J_3. Since J^2 will always commute with these, the effect of $\exp(i\theta n \cdot J)$ on a state $|j\ m>$ will be to produce an admixture of states $|j\ m'>$, but all having the same j.

In this example, we have considered SU(2) operations on a vector space which is the product of 2 vector spaces, each corresponding to the fundamental representation and hence, each 2-dimensional. So the initial vector space has 4 dimensions and the operators, which are tensors of the form $\sigma^a\sigma^b$, are also 4-dimensional. But the vector space has been reduced to two sets: a triplet with $j = 1$ and a singlet with $j = 0$. The **irreducible representations** which operate on these now are the 3-dimensional matrices of Section 4.3 and the unit operator in 1D.

5.3 Meson States

If we now interpret the states in terms of the u and d quark wavefunctions, we can build the states in terms of the isospin quantum numbers:

$$\begin{aligned}|1+1> &= uu \\ |10> &= \tfrac{1}{\sqrt{2}}(ud + du) \\ |1-1> &= dd \\ |00> &= \tfrac{1}{\sqrt{2}}(ud - du)\end{aligned} \qquad (5.29)$$

However, there are no observed qq states. Instead, if we wish to create the mesons, which are $q\bar{q}$ states, we recall that:

$$\bar{\mathbf{2}} = \begin{pmatrix} \bar{d} \\ -\bar{u} \end{pmatrix} \text{ transforms like } \mathbf{2} \equiv \begin{pmatrix} u \\ d \end{pmatrix}.$$

So to obtain the $q\bar{q}$ states, we make the replacements $u \to \bar{d}$ and $d \to -\bar{u}$:

$$
\begin{aligned}
|1+1> &= u\bar{d} \\
|10> &= (-)\tfrac{1}{\sqrt{2}}(u\bar{u} - d\bar{d}) \\
|1-1> &= (-)d\bar{u} \\
|00> &= \tfrac{1}{\sqrt{2}}(u\bar{u} + d\bar{d})
\end{aligned}
\tag{5.30}
$$

While some of the signs are important (those which are "internal"), there are global phases which are not significant and can be dropped.

We have a triplet with $I = 1$ and a singlet with $I = 0$ and we associate these states with the triplet of pions: $\{\pi^+, \pi^0, \pi^-\}$. But what about the singlet? Suitable, low-mass states include the η or the η' – but which is it? The identification of the singlet will have to wait until we look at $SU(3)_{\text{flavour}}$.

5.4 Weights

We want to find the set of commuting Hermitian operators for a given system, since these correspond to the independent, simultaneous observables. The set of commuting generators is called the **Cartan subalgebra**. The associated eigenvalues are the **weights**. **Weight vectors** are constructed from the eigenvalues from all of the commuting generators – there is one weight vector for each base vector. (The weights of the adjoint representation are called the **roots** – they are related to the raising and lowering operators.)

In SU(2), the simplest representation of the generators is in terms of the 2×2 Pauli σ matrices, of which there are $2^2 - 1 = 3$. The generators do not commute with each other, so the set of "commuting generators" is trivial, just consisting of a single generator. The generator J_3 is chosen, giving rise to 2 weight vectors, $(m = -^1/_2)$ and $(m = +^1/_2)$ – these are just points on a line. In SU(3), the simplest representation is in terms of the

3×3 λ matrices, of which there are $(3^2-1) = 8$ (see next chapter). There are 2 commuting generators, λ_3 and λ_8, giving rise to 3 weight vectors, of the form (m, Y) – these are points in a plane.

The weights are useful for classifying states and combinations of states.

5.4.1 *Weights in SU(2)*

As we just saw, the weights for the fundamental representation of SU(2) are $-^1/_2$ and $+^1/_2$. We can represent these as a pair of points on a line, as illustrated in Fig. 5.1.

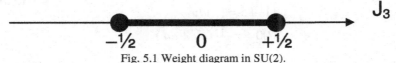

Fig. 5.1 Weight diagram in SU(2).

5.4.1.1 *Weights for meson states*

When we combine **two particles** (meson state or two spin-$^1/_2$ states), in the language of spin:

Spin $^1/_2$ $\otimes$ $^1/_2$ =   1 $\oplus$ 0
Multiplicity   2 $\times$ 2   =   3   +   1

The symbol $\otimes$ indicates a "combination", while $\oplus$ indicates an "addition". For two particles, a and b, we add the weight diagrams for the second particle b to the nodes of the first a, as illustrated in Fig. 5.2.

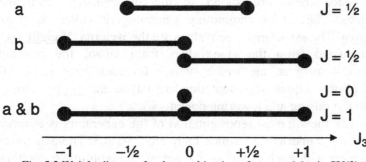

Fig. 5.2 Weight diagram for the combination of two particles in SU(2).

The result is a triplet and a singlet – the states identified in Eq. (5.30).

5.4.1.2 *Weights for baryon states*

When we combine **three particles** (baryon state or three spin-$^1/_2$ states), in the language of spin:

Spin $\qquad \qquad ^1/_2 \quad \otimes \quad ^1/_2 \quad \otimes \quad ^1/_2 \quad = \quad (1 \quad \oplus \quad 0) \quad \otimes \quad ^1/_2$

$\qquad \qquad \qquad \qquad \qquad \qquad \qquad \qquad = \quad ^3/_2 \quad \oplus \quad ^1/_2 \quad \oplus \quad ^1/_2$

Multiplicity $\qquad 2 \quad \times \quad 2 \quad \times \quad 2 \quad = \quad 4 \quad + \quad 2 \quad + \quad 2$

To the triplet and singlet already identified from combining two states, we add the weight diagram corresponding to the third state, as illustrated in Fig. 5.3.

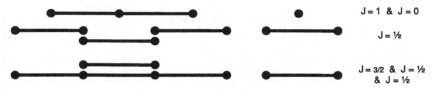

Fig. 5.3 Weight diagram for the combination of three particles in SU(2). The sets of diagrams associated with the triplet and singlet have been separated for clarity – the J_3 axis is no longer used; we just look at sets of nodes corresponding to multiplets

The result is a quadruplet and two doublets.

The weight diagrams help us to identify the quantum numbers of the resultant states – in what follows, we will use I to identify explicitly the isospin. To find the quark description of the states built from three particles, we proceed as with the mesons:

1. Construct the maximally aligned state: *uuu* corresponding to $I = {}^3/_2, I_3 = {}^3/_2$.
2. Use the lowering operator: $I_- = I_-{}^a + I_-{}^b + I_-{}^c$ to create states of lower I_3.
3. Construct other states of lower I using orthogonality and use the raising or lowering operators to complete the multiplets.

The last step is not unambiguous, but is usually done by starting from states of well-defined symmetry for two particles and adding a third, in a manner consistent with the construction of the weight diagrams in Fig. 5.3 – see also the next section on Glebsch-Gordon coefficients. This process results in the states shown in Table 5.1.

We can identify the $I = {}^3/_2$ multiplet as the particles $\{\Delta^{++}, \Delta^{+}, \Delta^{0}, \Delta^{-}\}$ But which states correspond to $\{p, n\}$, since there are two $I = {}^1/_2$ multiplets? Again, the answer to this will have to wait.

Table 5.1 Combination of 3 quarks under $SU(2)_{\text{flavour}}$. (Net signs are not significant.)

$I = {}^3/_2 (= 1 + {}^1/_2)$	Symmetric in $1 \leftrightarrow 2 \leftrightarrow 3$
$I_3 = +{}^3/_2$	uuu
$I_3 = +{}^1/_2$	$\frac{1}{\sqrt{3}}(uud + udu + duu)$
$I_3 = -{}^1/_2$	$\frac{1}{\sqrt{3}}(udd + dud + ddu)$
$I_3 = -{}^3/_2$	ddd
$I = {}^1/_2 (= 1 - {}^1/_2)$	Symmetric in $1 \leftrightarrow 2$
$I_3 = +{}^1/_2$	$\sqrt{\frac{2}{3}}(uud - \frac{udu + duu}{2}) \sim \frac{1}{\sqrt{6}}((ud + du)u - 2uud)$
$I_3 = -{}^1/_2$	$\sqrt{\frac{2}{3}}(ddu - \frac{dud + udd}{2}) \sim \frac{1}{\sqrt{6}}((du + ud)d - 2ddu)$
$I = {}^1/_2 (= 0 + {}^1/_2)$	Antisymmetric in $1 \leftrightarrow 2$
$I_3 = +{}^1/_2$	$\frac{1}{\sqrt{2}}(ud - du)u$
$I_3 = -{}^1/_2$	$\frac{1}{\sqrt{2}}(du - ud)d$

5.5 Clebsch–Gordon Coefficients

The coefficients multiplying the different contributions to the wavefunctions when combining states are the **Clebsch–Gordon coefficients**. They are defined by:

$$| jm > = \sum_{m_1, m_2} C_{m \, m_1 \, m_2}^{j \, j_1 \, j_2} | j_1 m_1 > | j_2 m_2 > \qquad (5.31)$$

with $m = m_1 + m_2$. These numbers are tabulated in the *PDG Book* [PDG] – their use is pretty self-explanatory.

For example, the $I = {}^3/_2$, $I_3 = {}^1/_2$ state can be constructed from the $I = 1$ and $I = {}^1/_2$ states. In turn, the $I = 1$ states can be expanded using the appropriate Clebsch–Gordon coefficients or by reference to Eq. (5.29):

$$\begin{pmatrix} \frac{3}{2} \\ \frac{+1}{2} \end{pmatrix} = \sqrt{\frac{1}{3}} \begin{pmatrix} 1 \\ +1 \end{pmatrix} \begin{pmatrix} \frac{1}{2} \\ \frac{-1}{2} \end{pmatrix} + \sqrt{\frac{2}{3}} \begin{pmatrix} 1 \\ 0 \end{pmatrix} \begin{pmatrix} \frac{1}{2} \\ \frac{+1}{2} \end{pmatrix}$$

$$= \sqrt{\frac{1}{3}}(uu)d + \sqrt{\frac{2}{3}} \frac{ud + du}{\sqrt{2}} u = \frac{1}{\sqrt{3}}(uud + udu + duu)$$

(5.32)

The same states can be used to build the $I = {}^1/_2$, $I_3 = {}^1/_2$ state:

$$\begin{pmatrix} \frac{1}{2} \\ \frac{+1}{2} \end{pmatrix} = \sqrt{\frac{2}{3}} \begin{pmatrix} 1 \\ +1 \end{pmatrix} \begin{pmatrix} \frac{1}{2} \\ \frac{-1}{2} \end{pmatrix} - \sqrt{\frac{1}{3}} \begin{pmatrix} 1 \\ 0 \end{pmatrix} \begin{pmatrix} \frac{1}{2} \\ \frac{+1}{2} \end{pmatrix} =$$

$$\sqrt{\frac{2}{3}}(uu)d - \sqrt{\frac{1}{3}} \frac{ud + du}{\sqrt{2}} u = \sqrt{\frac{2}{3}}(uud - \frac{1}{2}(ud + du)u)$$

(5.33)

$$\sim \frac{1}{\sqrt{6}}((ud + du)u - 2uud)$$

By construction, this is symmetric in $1 \leftrightarrow 2$.

What about the antisymmetric form in $1 \leftrightarrow 2$ made from the $I = 0$ and $I = {}^1/_2$ states? There is only one pair which will give the appropriate value of I_3, hence the description is trivial:

$$\begin{pmatrix} \frac{1}{2} \\ \frac{+1}{2} \end{pmatrix} = \begin{pmatrix} 0 \\ 0 \end{pmatrix} \begin{pmatrix} \frac{1}{2} \\ \frac{+1}{2} \end{pmatrix} = \frac{ud - du}{\sqrt{2}} u$$

(5.34)

Chapter 6

The Special Unitary Group SU(3)

6.1 Introduction

In this chapter, we will consider the special unitary group **SU(3)** and make connection to **quantum chromodynamics** (QCD). We will see how the multi-particle representations of SU(3) can be used to describe the **hadron wavefunctions**. Also we will see how these states can be described in terms of **Young tableaux**.

The key messages to extract from this chapter are:

- SU(3) provides a description of the exchange bosons (**gluons**) of **QCD** and allows the interactions of **coloured quarks** to be calculated.
- **Young tableaux** can be used to describe **multiplets**, allowing their **multiplicities** to be calculated.

6.2 The Gell-Mann Matrices

<u>SU(3)</u> corresponds to the set of special unitary transformations which act on complex 3D vectors and the operation of matrix multiplication. The natural representation is that of 3×3 matrices acting on complex 3D vectors. There are 3^2-1 parameters, hence 8 generators: $\{X_1, X_2, ..., X_8\}$. The generators are traceless and Hermitian. The generators are derived from the <u>**Gell-Mann matrices**</u>:

$$X_i = \tfrac{1}{2}\lambda_i \tag{6.1}$$

where:

$$\lambda_1 = \begin{pmatrix} 0 & 1 & 0 \\ 1 & 0 & 0 \\ 0 & 0 & 0 \end{pmatrix}, \quad \lambda_2 = \begin{pmatrix} 0 & -i & 0 \\ +i & 0 & 0 \\ 0 & 0 & 0 \end{pmatrix}, \quad \lambda_3 = \begin{pmatrix} 1 & 0 & 0 \\ 0 & -1 & 0 \\ 0 & 0 & 0 \end{pmatrix},$$

$$\lambda_4 = \begin{pmatrix} 0 & 0 & 1 \\ 0 & 0 & 0 \\ 1 & 0 & 0 \end{pmatrix}, \quad \lambda_5 = \begin{pmatrix} 0 & 0 & -i \\ 0 & 0 & 0 \\ +i & 0 & 0 \end{pmatrix} \tag{6.2}$$

$$\lambda_6 = \begin{pmatrix} 0 & 0 & 0 \\ 0 & 0 & 1 \\ 0 & 1 & 0 \end{pmatrix}, \quad \lambda_7 = \begin{pmatrix} 0 & 0 & 0 \\ 0 & 0 & -i \\ 0 & +i & 0 \end{pmatrix}, \quad \lambda_8 = \tfrac{1}{\sqrt{3}} \begin{pmatrix} 1 & 0 & 0 \\ 0 & 1 & 0 \\ 0 & 0 & -2 \end{pmatrix}$$

However, note that:

$$\lambda_0 = \begin{pmatrix} 1 & 0 & 0 \\ 0 & 1 & 0 \\ 0 & 0 & 1 \end{pmatrix} \tag{6.3}$$

is not part of SU(3) – it corresponds to a U(1):

$$U(3) = SU(3) \otimes U(1) \tag{6.4}$$

By construction there is an obvious SU(2) subgroup provided by the associations $\lambda_{1,2,3} \leftrightarrow \sigma_{1,2,3}$. While $\sigma_{1,2}$ have a role in forming raising and lowering operators, so will the pairs $\lambda_{1,2}$, $\lambda_{4,5}$ and $\lambda_{6,7}$.

The matrices are chosen to satisfy:

$$\mathrm{Tr}(\lambda_i \lambda_j) = 2\delta_{ij} \tag{6.5}$$

The structure constants are defined by:

$$[X_i, X_j] = i\sum_k f_{ijk} X_k \tag{6.6}$$

and are non-trivial; they can be found in the *PDG Book* [PDG].

6.3 Quantum Chromodynamics (QCD)

QCD is associated with the colour charges of the quarks and gluons. We choose base vectors:

Red: r or $|r>$ $= (1, 0, 0)$

Blue: b or $|b>$ $= (0, 1, 0)$

Green: g or $|g>$ $= (0, 0, 1)$

I will flip between notations quite cavalierly.

By construction the Standard Model has an exact $SU(3)_{colour}$ local symmetry, with a corresponding gauge invariance and 8 associated gauge bosons – **gluons**.

The gauge-invariant term which can be included in the Lagrangian for the gauge fields G_i is:

$$L_{gauge} \sim F_{i\,\mu\nu} F_i^{\,\mu\nu} \tag{6.7}$$

This represents the energy stored in the fields. $F_{i\,\mu\nu}$ is derived from the commutator of the covariant derivatives:

$$F_i^{\,\mu\nu} X_i \sim [D^\mu, D^\nu] \sim (\partial^\mu G_i^{\,\nu} - \partial^\nu G_i^{\,\mu} + f_{ijk} G_j^{\,\mu} G_k^{\,\nu}) X_i \tag{6.8}$$

where:

$$D^\mu = \partial^\mu - iX_i G_i^{\,\mu} \tag{6.9}$$

and where there is an implicit sum over the 8 gluon fields: $i = 1, ..., 8$. (For $U(1)_{EM}$, $F_{\mu\nu}$ is the field tensor, corresponding to the electric and magnetic fields.)

For a non-Abelian theory like $SU(3)_{colour}$, the structure constants f_{ijk} are non-vanishing and there are terms in L_{gauge} which correspond to triple and quartic gauge couplings, i.e. the gluons couple to themselves. These types of interactions are also apparent in $SU(2)_{weak\ isospin}$, leading to interactions between the W bosons. However, since $U(1)_{QED}$ is an Abelian theory, there are no interactions between photons.

Due to a conspiracy of the QCD couplings (arising from the properties of SU(3)), the energy involved in separating two coloured charges is infinite. Therefore, free observable particles must correspond to states which have no colour – they are "colourless".

To understand what a "colourless" state is, we recall the formulation of a gauge theory, whereby transformations are associated with a "charge":

$$U = \exp(iaX) \tag{6.10}$$

The generator projects out the charge of the state, to be multiplied by the unit of "charge" a. If the charge of the state is zero, the effect of the transformation will be to leave the state unchanged – it transforms to itself and is thus described as a "singlet" (as opposed to a multiplet, where sets of states transform amongst themselves) – see the example in Section 5.2.1. So we need to identify states which are left invariant by all possible gauge transformations.

6.3.1 *The colour of hadronic states*

For the description of the **baryon** colour wavefunction in terms of the colours of the three individual quarks, we need to construct invariant states which are "colourless". It is tempting to consider $|r>|b>|g>$, however this is not colourless. For example under the infinitesimal transformation:

$$U = \exp(i\alpha\lambda_1) = 1 + i\alpha\lambda_1 \tag{6.11}$$

each individual quark wavefunction transforms separately[a] and the net effect is that the combined state transforms as follows:

$$
\begin{aligned}
|r>|b>|g> &\to (|r> + i\alpha\,|b>)(|b> + i\alpha\,|r>)|g> \\
&= |r>|b>|g> + i\alpha\,|b>|b>|g> + i\alpha\,|r>|r>|g> \\
&\neq |r>|b>|g>
\end{aligned} \tag{6.12}
$$

---

[a] Consider the transformation of the base vectors given at the start of Section 6.3 by the λ matrices.

Instead we construct a (tensor) state from a linear combination:

$$\psi = \sum_{ijk} c_{ijk} \, |i>|j>|k> \tag{6.13}$$

where i, j, k are taken from the set $\{r, b, g\}$. Under a unitary transformation U, each of the individual quark states needs to be transformed separately:

$$\psi' = \sum_{ijk} c_{ijk} U(|i>)U(|j>)U(|k>) \tag{6.14}$$

Evaluating the effects of the operators in terms of the matrix components:

$$U\,|i> = \sum_p |p><p|U\,|i> = \sum_p U_{pi}\,|p> \tag{6.15}$$

we find:

$$\psi' = \sum_{ijk} \sum_{pqr} c_{ijk} U_{pi} U_{qj} U_{rk} \, |p>|q>|r> \tag{6.16}$$

If we chose $c_{ijk} = \varepsilon_{ijk}$, then:

$$\begin{aligned}
\psi' &= \sum_{ijk}\sum_{pqr} \varepsilon_{ijk} U_{pi} U_{qj} U_{rk} \, |p>|q>|r> \\
&= \sum_{pqr} \varepsilon_{pqr} \det(U)\, |p>|q>|r> \\
&= \sum_{pqr} \varepsilon_{pqr} \, |p>|q>|r> = \psi
\end{aligned} \tag{6.17}$$

where we have used the well-known expression for the determinant of a matrix. So we have successfully identified a singlet of the transformation and hence the colour description of a baryon is:

$$\begin{aligned}
&|r>|b>|g> + |g>|r>|b> + |b>|g>|r> - \\
&|r>|g>|b> - |b>|r>|g> - |g>|b>|r>
\end{aligned} \tag{6.18}$$

or dropping the ket notation:

$$\psi_{\text{baryon}} \sim rbg + grb + bgr - rgb - brg - gbr \qquad (6.19)$$

Now we consider the description of a **meson**. The conjugate state transforms like:

$$\overline{\psi} \to \overline{\psi}' = \overline{\psi} U^H \qquad (6.20)$$

and is represented by the bra. So for example:

$$< r | \to < r' | = < r | U^H \qquad (6.21)$$

For the meson colour wavefunction, it is tempting to consider something like $| r >< r |$. However, just as before, this is not colourless. For example under the infinitesimal transformation of Eq. (6.11):

$$| r >< r | \to (| r > + i\alpha | b >)(< r | - i\alpha < b |)$$
$$= | r >< r | - i\alpha | r >< b | + i\alpha | b >< r | \qquad (6.22)$$
$$\neq | r >< r |$$

Instead, in a similar fashion to what we did for the baryons, we construct a (tensor) state from a linear combination:

$$\psi = \sum_{ij} c_{ij} | i >< j | \qquad (6.23)$$

Under a unitary transformation, U:

$$\psi' = \sum_{ij} c_{ij} U | i >< j | U^H \qquad (6.24)$$

Expanding as before:

$$\psi' = \sum_{ij} \sum_{pq} c_{ij} | p >< p | U | i >< j | U^H | q >< q | \qquad (6.25)$$

If we chose $c_{ij} = \delta_{ij}$, then:

$$\psi' = \sum_{ij}\sum_{pq} \delta_{ij} \mid p><p \mid U \mid i><j \mid U^H \mid q><q \mid$$

$$= \sum_{i}\sum_{pq} \mid p><p \mid U \mid i><i \mid U^H \mid q><q \mid$$

$$= \sum_{pq} \mid p><p \mid UU^H \mid q><q \mid \qquad (6.26)$$

$$= \sum_{pq} \mid p>\delta_{pq}<q \mid$$

$$= \sum_{p} \mid p><p \mid = \psi$$

So we have successfully identified a singlet of the transformation and hence the colour description of a meson is:

$$\mid r><r \mid + \mid b><b \mid + \mid g><g \mid \qquad (6.27)$$

or:

$$\psi_{\text{meson}} \sim r\bar{r} + b\bar{b} + g\bar{g} \qquad (6.28)$$

It is easy to show that the effect of the generators on these states results in eigenvalues of zero, i.e. the states are indeed colourless.[b]

6.3.2 *Gluons*

Gluon fields $\{G_i: i=1, ..., 8\}$ are required to ensure the invariance of the fermionic terms in the SM Lagrangian:

$$L \sim \bar{\psi}\!\!\!D\psi \quad \text{where} \quad D \sim \partial - i\tfrac{1}{2}\lambda_i G_i \qquad (6.29)$$

---

[b] The fact that *red* + *blue* + *green* light appears to make white light is purely a feature of the physiology of the human eye and the fact that the cones are sensitive to red, blue and green light. The analogy between visible light and the quantum numbers of the SU(3) group is a convenience, nothing more.

This gives terms in the Lagrangian like:

$$\bar{\psi}\lambda_1 G_1 \psi = \begin{pmatrix} \bar{r} & \bar{b} & \bar{g} \end{pmatrix} \begin{pmatrix} 0 & 1 & 0 \\ 1 & 0 & 0 \\ 0 & 0 & 0 \end{pmatrix} G_1 \begin{pmatrix} r \\ b \\ g \end{pmatrix} = \bar{r} G_1 b + \bar{b} G_1 r \qquad (6.30)$$

where b really corresponds to a creation operator for a blue state $|b>$. We interpret these labels as operators which can operate on the vacuum. Let us look at the first term, illustrated in Fig. 6.1.

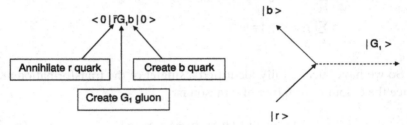

Fig. 6.1 Gluon vertex.

In this example, we deduce $G_1 \sim r\bar{b}$ as far as the colour quantum numbers are concerned. So looking at both terms in Eq. (6.30), we have:

$$G_1 \sim \frac{1}{\sqrt{2}}(r\bar{b} + b\bar{r}) \qquad (6.31)$$

Likewise, by considering the other terms in the Lagrangian, we can identify all 8 coloured gluons associated with SU(3)$_{\text{QCD}}$. These are shown in Table 6.1.

Table 6.1 Colour wavefunctions for the 8 gluons.

$G_1 \sim \frac{1}{\sqrt{2}}(r\bar{b} + b\bar{r})$	$G_2 \sim \frac{i}{\sqrt{2}}(r\bar{b} - b\bar{r})$	$G_3 \sim \frac{1}{\sqrt{2}}(r\bar{r} - b\bar{b})$
$G_4 \sim \frac{1}{\sqrt{2}}(b\bar{g} + g\bar{b})$	$G_5 \sim \frac{i}{\sqrt{2}}(b\bar{g} - g\bar{b})$	
$G_6 \sim \frac{1}{\sqrt{2}}(g\bar{r} + r\bar{g})$	$G_7 \sim \frac{i}{\sqrt{2}}(g\bar{r} - r\bar{g})$	$G_8 \sim \frac{1}{\sqrt{6}}(r\bar{r} + b\bar{b} - 2g\bar{g})$

The coefficients from the Lagrangian contained in the Gell-Mann matrices are absorbed into the description of the gluon wavefunctions.

The singlet:

$$\tfrac{1}{\sqrt{3}}(r\bar{r} + b\bar{b} + g\bar{g}) \tag{6.32}$$

corresponding to a U(1) transformation, is not an "observed" gluon state.

In analogy with the charged raising and lowering operators we identified in $SU(2)_{\text{weak isospin}}$:

$$W^\pm = \tfrac{1}{\sqrt{2}}(W_1 \mp iW_2) \tag{6.33}$$

the gluon states can be combined to create "charge operators":

$$r\bar{b} = \tfrac{1}{\sqrt{2}}(G_1 - iG_2) \quad \text{etc.} \tag{6.34}$$

These represent the flow of colour "charge", corresponding to the exchange of gluons from one quark to another and will be more useful in what follows.

6.3.3 Colour factors

The **scattering amplitude** for the process illustrated in Fig. 6.2 is proportional to the couplings at the vertices: $g_1 g_2$. These coupling strengths are found from the projection of the initial and final colour

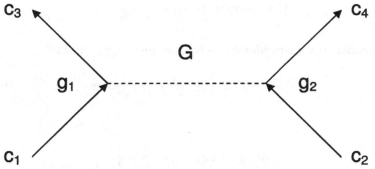

Fig. 6.2 Feynman diagram for gluon exchange between two coloured quarks.

states $|c_i\rangle|c_f\rangle$ on to the gluon state $|G\rangle$. According to the rules of quantum field theory, we include a $-$ve sign for antiquarks. In what follows, we will evaluate the colour factors which correspond to the coloured parts of the scattering amplitudes for all possible combinations of coloured quarks. This is illustrated in Figs 6.3–6.7.

Exchange gluons:

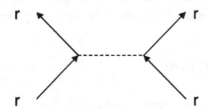

$$G_3 \sim \tfrac{1}{\sqrt{2}}(r\bar{r} - b\bar{b})$$
$$G_8 \sim \tfrac{1}{\sqrt{6}}(r\bar{r} + b\bar{b} - 2g\bar{g})$$

Fig. 6.3 Feynman diagram for gluon exchange between **similar quarks:** $qq \rightarrow qq$, e.g. $rr \rightarrow rr$.

For the process in Fig. 6.3, the amplitudes are:

$$A(rr \rightarrow rr) = \tfrac{1}{\sqrt{2}} \cdot \tfrac{1}{\sqrt{2}} \quad \text{via } G_3 \qquad (6.35)$$

and:

$$A(rr \rightarrow rr) = \tfrac{1}{\sqrt{6}} \cdot \tfrac{1}{\sqrt{6}} \quad \text{via } G_8 \qquad (6.36)$$

So:

$$A(rr \rightarrow rr) = \tfrac{1}{\sqrt{2}} \cdot \tfrac{1}{\sqrt{2}} + \tfrac{1}{\sqrt{6}} \cdot \tfrac{1}{\sqrt{6}} = \tfrac{2}{3} \qquad (6.37)$$

Likewise, if we consider $bb \rightarrow bb$ and $gg \rightarrow gg$, we find:

$$A(bb \rightarrow bb) = \tfrac{-1}{\sqrt{2}} \cdot \tfrac{-1}{\sqrt{2}} + \tfrac{1}{\sqrt{6}} \cdot \tfrac{1}{\sqrt{6}} = \tfrac{2}{3} \qquad (6.38)$$

and:

$$A(gg \rightarrow gg) = \tfrac{-2}{\sqrt{6}} \cdot \tfrac{-2}{\sqrt{6}} = \tfrac{2}{3} \qquad (6.39)$$

So the amplitudes for all 3 colours are the same – there is invariance to the colour of the quark, as one would expect if there is to be colour symmetry. So:

$$A(qq \to qq) = \tfrac{2}{3} \qquad (6.40)$$

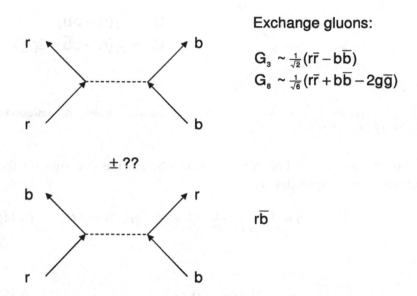

Exchange gluons:

$$G_3 \sim \tfrac{1}{\sqrt{2}}(r\bar{r} - b\bar{b})$$
$$G_8 \sim \tfrac{1}{\sqrt{6}}(r\bar{r} + b\bar{b} - 2g\bar{g})$$

$r\bar{b}$

Fig. 6.4 Feynman diagram for gluon exchange between **different quarks:** $qq' \to qq'$, e.g. $rb \to rb$. The combination of the amplitudes depends on the symmetry of the initial state.

For the process in Fig. 6.4, there are two amplitudes. Whether we add or subtract them depends on the exchange symmetry of the wavefunction describing the incoming quarks.

The two amplitudes are:

$$A(rb \to rb) = \tfrac{1}{\sqrt{2}} \cdot \tfrac{-1}{\sqrt{2}} + \tfrac{1}{\sqrt{6}} \cdot \tfrac{1}{\sqrt{6}} = \tfrac{-1}{3} \quad \text{via } G_3 \text{ and } G_8 \qquad (6.41)$$

and:

$$A(rb \to br) = 1 \cdot 1 = 1 \quad \text{via } r\bar{b} \qquad (6.42)$$

So:

$$A(qq' \rightarrow qq') = \tfrac{-1}{3} \pm 1 \qquad (6.43)$$

depending on the symmetry.

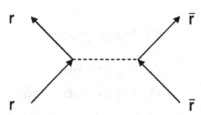

Exchange gluons:

$$G_3 \sim \tfrac{1}{\sqrt{2}}(r\bar{r} - b\bar{b})$$
$$G_8 \sim \tfrac{1}{\sqrt{6}}(r\bar{r} + b\bar{b} - 2g\bar{g})$$

Fig. 6.5 Feynman diagram for gluon exchange between **quark & antiquark:** $q\bar{q} \rightarrow q\bar{q}$, e.g. $r\bar{r} \rightarrow r\bar{r}$.

For the process in Fig. 6.5, and remembering the –ve sign for the antiquarks, the amplitudes are:

$$A(r\bar{r} \rightarrow r\bar{r}) = \tfrac{1}{\sqrt{2}} \cdot \tfrac{-1}{\sqrt{2}} + \tfrac{1}{\sqrt{6}} \cdot \tfrac{-1}{\sqrt{6}} = \tfrac{-2}{3} \quad \text{via } G_3 \text{ and } G_8 \qquad (6.44)$$

So:

$$A(q\bar{q} \rightarrow q\bar{q}) = \tfrac{-2}{3} \qquad (6.45)$$

Note: we have ignored the *s*-channel scattering.

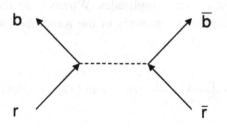

Exchange gluons:

$$r\bar{b}$$

Fig. 6.6 Feynman diagram for gluon exchange between **quark & antiquark:** $q\bar{q} \rightarrow q'\bar{q}'$, e.g. $r\bar{r} \rightarrow b\bar{b}$.

For the process in Fig. 6.6, the amplitude is:

$$A(r\bar{r} \to b\bar{b}) = 1 \cdot -1 = -1 \quad \text{via } r\bar{b} \tag{6.46}$$

So:

$$A(q\bar{q} \to q'\bar{q}') = -1 \tag{6.47}$$

Again the s-channel scattering has been ignored.

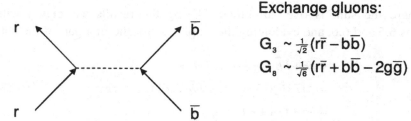

Exchange gluons:

$$G_3 \sim \tfrac{1}{\sqrt{2}}(r\bar{r} - b\bar{b})$$
$$G_8 \sim \tfrac{1}{\sqrt{6}}(r\bar{r} + b\bar{b} - 2g\bar{g})$$

Fig. 6.7 Feynman diagram for gluon exchange between **quark & antiquark:** $q\bar{q}' \to q\bar{q}'$, e.g. $r\bar{b} \to r\bar{b}$.

For the process in Fig. 6.7, the amplitudes are:

$$A(r\bar{b} \to r\bar{b}) = \tfrac{1}{\sqrt{2}} \cdot \tfrac{+1}{\sqrt{2}} + \tfrac{1}{\sqrt{6}} \cdot \tfrac{-1}{\sqrt{6}} = \tfrac{1}{3} \quad \text{via } G_3 \text{ and } G_8 \tag{6.48}$$

So:

$$A(q\bar{q}' \to q\bar{q}') = \tfrac{1}{3} \tag{6.49}$$

6.3.4 Example – using colour factors

This is a very simplistic example, where we consider the gluon interactions between quarks within a meson, as illustrated in Fig. 6.8.

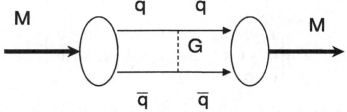

Fig. 6.8 Feynman diagram for gluon exchange between quarks within a meson.

The description of the meson is as a colour singlet:

$$|M> = \frac{1}{\sqrt{3}} |r\bar{r} + b\bar{b} + g\bar{g}> \qquad (6.50)$$

The amplitude describing the propagation of the meson is:

$$<M|\sum G|M> \qquad (6.51)$$

where the sum is over all gluons. Using the results associated with Figs 6.5 and 6.6, and evaluating the amplitude for the first part of the ket:

$$< r\bar{r} + b\bar{b} + g\bar{g} | \sum G | r\bar{r} >$$
$$= < r\bar{r} | \sum G | r\bar{r} > + < b\bar{b} + g\bar{g} | \sum G | r\bar{r} > \qquad (6.52)$$
$$= \frac{-2}{3} + (-1 + -1) = \frac{-8}{3}$$

Recalling the normalisation and anticipating that the amplitudes for $|b\bar{b}>$ and $|g\bar{g}>$ will be the same as $|r\bar{r}>$ by symmetry, the net amplitude is:

$$\sim 3 \times (\frac{1}{\sqrt{3}})^2 \times \frac{-8}{3} = \frac{-8}{3} \qquad (6.53)$$

Expressed like this, the value is not very meaningful, but it illustrates how calculations can be done.

6.4 Weights for SU(3)

The commuting generators in SU(3) (Cartan subalgebra – see Chapter 5) are:

$$\lambda_3 = \begin{pmatrix} 1 & 0 & 0 \\ 0 & -1 & 0 \\ 0 & 0 & 0 \end{pmatrix} \quad \text{and} \quad \lambda_8 = \frac{1}{\sqrt{3}} \begin{pmatrix} 1 & 0 & 0 \\ 0 & 1 & 0 \\ 0 & 0 & -2 \end{pmatrix} \qquad (6.54)$$

This implies there are two simultaneously observable quantum numbers, along with I^2. With an eye to the description of hadrons, rather than QCD, we define:

$$\text{Isospin} \quad I_3 = \tfrac{1}{2}\lambda_3 \tag{6.55}$$

$$\text{Hypercharge} \quad Y = \frac{1}{\sqrt{3}}\lambda_8 \tag{6.56}$$

The weights are readily identified from the diagonal matrices:

$$I_3 = \tfrac{+1}{2}, \tfrac{-1}{2}, 0 \quad \text{and} \quad Y = \tfrac{1}{3}, \tfrac{1}{3}, \tfrac{-2}{3} \tag{6.57}$$

Just as for SU(2), where we defined raising and lowering operators which move between the different weight vectors (in SU(2), points on line), so we can define raising and lowering operators which move between the different weight vectors, which in SU(3), will be points in the plane:

$$I_\pm = \tfrac{1}{2}(\lambda_1 \pm i\lambda_2) \quad U_\pm = \tfrac{1}{2}(\lambda_6 \pm i\lambda_7) \quad V_\pm = \tfrac{1}{2}(\lambda_4 \pm i\lambda_5) \tag{6.58}$$

These are not all independent. We will consider these more in the next sections.

6.5 Quark Flavour in SU(3)

The quarks (u, d, s) are all light (compared to hadron masses) and their interactions are dominated by the flavour-independent colour force. We choose base vectors:

$$u = \begin{pmatrix} 1 \\ 0 \\ 0 \end{pmatrix}, \quad d = \begin{pmatrix} 0 \\ 1 \\ 0 \end{pmatrix}, \quad s = \begin{pmatrix} 0 \\ 0 \\ 1 \end{pmatrix} \tag{6.59}$$

The weights are:

	I_3	Y
u	$+^1/_2$	$+^1/_3$
d	$-^1/_2$	$+^1/_3$
s	0	$-^2/_3$

Just as:

$$I_+ = \begin{pmatrix} 0 & 1 & 0 \\ 0 & 0 & 0 \\ 0 & 0 & 0 \end{pmatrix} \tag{6.60}$$

raises $d = (0, 1, 0)$ to $u = (1, 0, 0)$, so:

$$U_+ = \begin{pmatrix} 0 & 0 & 0 \\ 0 & 0 & 1 \\ 0 & 0 & 0 \end{pmatrix} \tag{6.61}$$

raises $s = (0, 0, 1)$ to $d = (0, 1, 0)$.

The weight diagram is shown in Fig. 6.9.

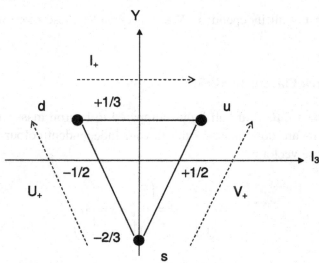

Fig. 6.9 Weight diagram for light quarks in $SU(3)_{\text{flavour}}$.

We simplify this by dropping all the labels. The results are the diagrams shown in Fig 6.10, which correspond to the fundamental representation for the quarks (**3**) and the antiquarks (**3̄**) (negated quantum numbers).

Fig. 6.10 Simplified weight diagrams for a quark and antiquark in SU(3).

To identify the states and multiplets for combining quarks and antiquarks, we must combine their diagrams. This is done by positioning a copy of the second (antiquark) diagram on top of each node in the first (quark) diagram. This corresponds to the vector addition of the quantum numbers of the particles. For example, Figs 6.11 and 6.12 show how two quarks or a quark and antiquark can be combined.

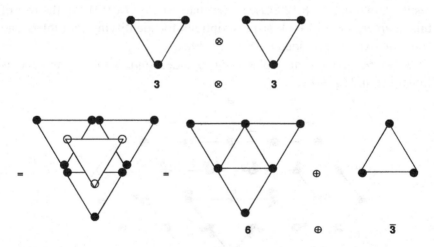

Fig. 6.11 Weight diagrams for the combination of two quarks in SU(3).

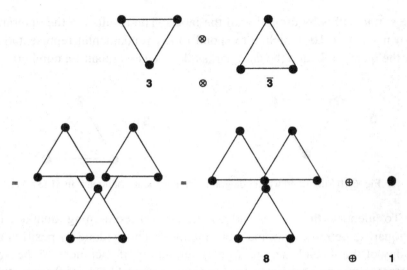

Fig. 6.12 Weight diagrams for the combination of a quark and an antiquark in SU(3).

These diagrams show the weights (quantum numbers), but how does one identify the multiplets associated with the symmetry? Why is the result shown in Fig. 6.12 8⊕1, rather than 7⊕2 or 7⊕1⊕1? In the rest of this chapter, we will look at prescriptions for identifying multiplets and some motivation provided by Young tableaux.

A more general multiplet with 6 sides and 3-fold symmetry is illustrated in Fig. 6.13.

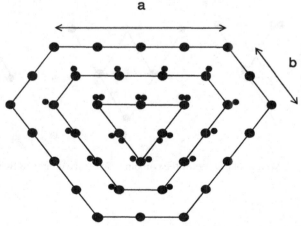

Fig. 6.13 More general weight diagrams for an SU(3) multiplet.

The rules for the construction of the diagram are:

i)   On the outer-most ring, there is only 1 state.
ii) On each subsequent inner ring, an extra state is added at each node.
iii) This continues until a triangle is obtained.
iv) After this, all inner triangles have the same number of states.

More detail can be found in [Georgi]. These rules can be used to verify the identification of the multiplets in Figs 6.11 and 6.12.

The multiplicity of the multiplet in Fig. 6.13 is given by:

$$\tfrac{1}{2}(a+1)(b+1)(a+b+2) \qquad (6.62)$$

This is tedious to prove, although not intrinsically difficult. So for example, the multiplet represented by Fig. 6.14:

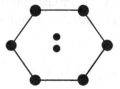

Fig. 6.14 An octet.

has $a = b = 1$, so the multiplicity is $^1/_2 \times 2 \times 2 \times 4 = 8$ – correct. Having tried to be more general, the most complex multiplets we ever get to worry about are for 3 quarks or antiquarks. So in $SU(3)_{\text{flavour}}$, the largest multiplicity corresponds to the decuplet shown in Fig. 6.15.

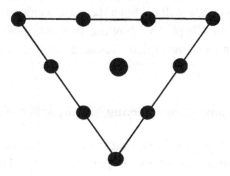

Fig. 6.15 A decuplet.

This has $a = 3$ and $b = 0$, so the multiplicity is $\frac{1}{2} \times 4 \times 1 \times 5 = 10$ – correct.

6.5.1 *Multi-particle states*

Eigenstates describing the combination of individual particles which are identical tend to have a well-defined exchange symmetry. The unitary symmetry operators will not change the **exchange symmetry** – the symmetry operators commute with the exchange operator. If we define an operator:

$$U = \exp(iaX^{ab}) \tag{6.63}$$

where $X^{ab} = X^a + X^b$ and X^a operates on particle a, and P^{ab} is the exchange operator such that particles a and b swap quantum numbers. Then:

$$\begin{aligned} P^{ab}U &= P^{ab}\exp(ia(X^a + X^b)) \\ &= \exp(ia(X^b + X^a))P^{ab} = \exp(ia(X^a + X^b))P^{ab} \\ &= UP^{ab} \end{aligned} \tag{6.64}$$

This means that the symmetry operations do not modify the exchange symmetry of a state. Since the multiplets consist of those states related to each other by unitary transformations (in particular, the raising and lowering operators), they will all have the same symmetry. We explicitly demonstrated this in Chapter 5, showing that the $I = 0$ and $I = 1$ states transformed within their multiplets. These multiplets are the irreducible representations.

6.6 Young Tableaux – Constructing Multi-particle States

Let us consider a 2-particle wavefunction: ψ^{ab} – where the index a (b) describes the quantum numbers of particle a (b). The exchange (or permutation) operator can be used to generate states of explicit symmetry:

$$S^{ab} = 1 + P^{ab} \quad \text{symmetrising operator} \tag{6.65}$$

$$A^{ab} = 1 - P^{ab} \quad \text{antisymmetrising operator} \tag{6.66}$$

Since:

$$P^{ab} P^{ab} = 1 \tag{6.67}$$

then:

$$P^{ab} S^{ab} = P^{ab} + 1 = S^{ab} \tag{6.68}$$

and:

$$P^{ab} A^{ab} = P^{ab} - 1 = -A^{ab} \tag{6.69}$$

So:

$$P^{ab}(S^{ab}\psi^{ab}) = +S^{ab}\psi^{ab} \tag{6.70}$$

and:

$$P^{ab}(A^{ab}\psi^{ab}) = -A^{ab}\psi^{ab} \tag{6.71}$$

This means that starting from a particular multi-particle state, we can apply S^{ab} and A^{ab} repeatedly to build up states which under exchange are:
- Symmetric with respect to all particles
- Antisymmetric with respect to all particles
- Mixed symmetry, which may be symmetric (antisymmetric) with respect to particular particles.

Rather than using "ψ", on to which we hang quantum-number labels, we use:

$\square$ to denote a particle – of which there are N.

"i" to label the state of a particle – of which there are n.

So the complete state for N particles is denoted by an arrangement of N boxes, each with its own state label. These diagrams are called **Young tableaux**, as illustrated in Fig. 6.16.

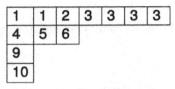

Fig. 6.16 Example of a Young tableau, with the state labels indicated.

The numbers in the boxes (could be labelled a, b, c, ... or u, d, s, ..., etc.) must not exceed n. The states are symmetrised with respect to all the particles ($\square$'s) in a given *row* and the states are antisymmetrised with respect to all the particles in a given *column*.

The rules for constructing the tableaux are:
1. A row must not be longer than the one above it.
2. The numbers (state labels) when viewed in reading order through the table must not decrease.
3. Going down vertically in a given column, the numbers must increase.

The rules ensure that we don't:
- double count (2nd rule)
- antisymmetrise with respect to the same single particle state, causing a vanishing combination (3rd rule)

6.6.1 Example – 2-particle states in SU(3)

Consider a 2-particle state: $N = 2$ and so we have two $\square$'s. In SU(3)$_{flavour}$, there are 3 labels ($n = 3$), e.g. (u, d, s) – we will call them generically (1, 2, 3). We can construct 6 symmetric states, as shown in Fig. 6.17 and 3 antisymmetric states, as shown in Fig. 6.18.

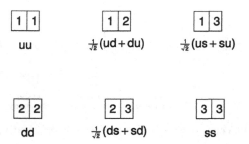

Fig. 6.17 Young tableaux for two particles under SU(3) – symmetric states.

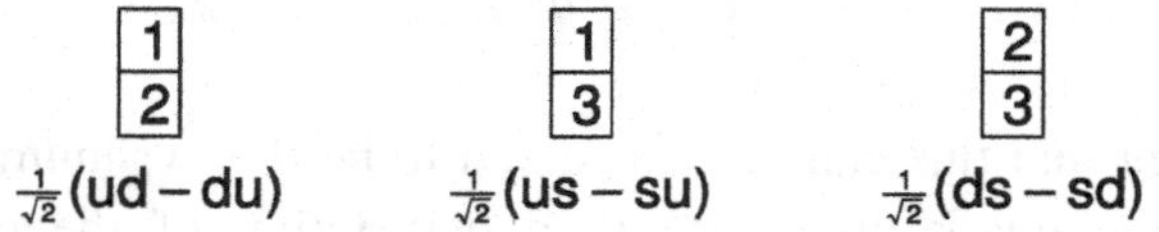

$$\boxed{\begin{array}{c}1\\2\end{array}} \qquad \boxed{\begin{array}{c}1\\3\end{array}} \qquad \boxed{\begin{array}{c}2\\3\end{array}}$$

$\frac{1}{\sqrt{2}}(\text{ud} - \text{du}) \qquad \frac{1}{\sqrt{2}}(\text{us} - \text{su}) \qquad \frac{1}{\sqrt{2}}(\text{ds} - \text{sd})$

Fig. 6.18 Young tableaux for two particles under SU(3) – antisymmetric states.

We have already seen the states:

$$uu, \quad \frac{1}{\sqrt{2}}(ud + du), \quad dd, \quad \frac{1}{\sqrt{2}}(ud - du) \qquad (6.72)$$

when considering the SU(2) subgroup.

The shapes of the tableaux correspond to the multiplets of the representations. Having motivated the Young tableaux, we will drop the state labels.

6.6.2 *Combining multiplets*

We will only be interested in fairly simple combinations of tableaux; hence, crudely speaking, it is sufficient to simply combine diagrams in a manner consistent with the rules given earlier. (The combination of more complex tableaux is explained in [Georgi].) For example, the combination of three particles is derived from that for two, as shown in Fig. 6.19.

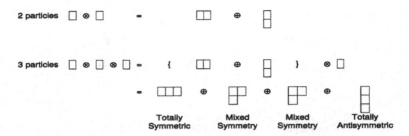

Fig. 6.19 Young tableaux for the combinations of 2 and 3 particles.

We recall that a totally antisymmetric singlet could be generated by:

$$\varepsilon_{ijk...}q^i q^j q^k ...$$ (6.73)

To the extent that this can be considered to be the "vacuum" state, then removing one quark gives rise to a description of the conjugate or antiparticle state:

$$\varepsilon_{ijk...}q^i q^j ...$$ (6.74)

So in SU(n), the corresponding Young tableau can be represented by a column of $n-1$ □'s, as shown in Fig. 6.20.

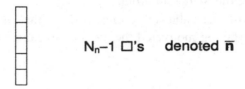

Fig. 6.20 Young tableau for an antiquark state.

Note that in SU(2), the conjugate is $\overline{\mathbf{2}} = \square$ ... which is the same as the quark state $\mathbf{2} = \square$. We know this because we showed that the conjugate state:

$$\overline{\mathbf{2}} = \begin{pmatrix} \overline{d} \\ -\overline{u} \end{pmatrix}$$ (6.75)

transforms like:

$$2 \equiv \begin{pmatrix} u \\ d \end{pmatrix} \tag{6.76}$$

The tableaux for the combination of a quark and an antiquark under SU(n) are shown in Fig. 6.21. The first multiplet is a singlet in SU(n).

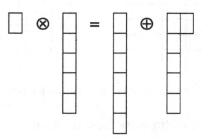

Fig. 6.21 Young tableaux for the combination of a quark and an antiquark.

6.6.3 *Calculating multiplicities*

The beauty of the Young tableaux is that they help us identify multiplets, understand their symmetry and evaluate their multiplicities. The multiplicity is the ratio of a numerator and a denominator. For the *numerator* in SU(n), we insert numbers into the cells of the tableau, as illustrated in Fig. 6.22:

- n along the diagonal.
- Decrease by 1 each step away from the diagonal, going down and left.
- Increase by 1 each step away from the diagonal, going up and right.

The numerator is evaluated from the product of all the numbers.

n	n+1	n+2	n+3	n+4	n+5	n+6
n−1	n	n+1				
n−2						
n−3						

Fig. 6.22 Evaluating the numerator for a tableau.

For the *denominator*, we evaluate the lengths of the "hooks". For each cell, a line is drawn starting on the right-hand side, entering the cell in question, and then leaving vertically downwards, as illustrated in Fig. 6.23. The length is the number of cells crossed. The denominator is evaluated from the product of all the numbers.

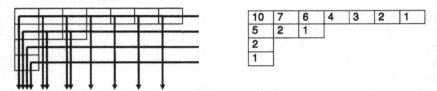

Fig. 6.23 Evaluating the denominator for a tableau.

Fig. 6.24 gives the multiplicity for several simple tableaux.

		numerator	denominator	multiplicity
n	$\boxed{n}$	n	1	n
$\bar{n}$	$\boxed{\begin{matrix} n \\ n{-}1 \\ ... \\ 2 \end{matrix}}$	$n(n{-}1) \ldots 2$	$(n{-}1) \ldots 1$	n
1	$\boxed{\begin{matrix} n \\ n{-}1 \\ ... \\ 1 \end{matrix}}$	$n!$	$n!$	1
	$\boxed{\begin{matrix} n & n{+}1 \\ n{-}1 \end{matrix}}$	$n(n{+}1)(n{-}1)$	$3{\cdot}1{\cdot}1$	$\dfrac{(n-1)n(n+1)}{3}$

E.g. $\quad SU(2) \rightarrow 2$
$\quad\quad\ \, SU(3) \rightarrow 8$

Fig. 6.24 Evaluating the multiplicities for simple tableaux.

Figs 6.25 and 6.26 show how the combinations of quarks lead to various multiplets. The multiplicities can be verified using the recipe given above. For different dimensions of symmetry groups, SU(n), the tableaux are identical when dealing with <u>quarks</u>, albeit that the multiplicities are different. However, since the representation of the antiquark depends on n, when antiquarks are involved, the tableaux are different for different n.

3 Quarks:

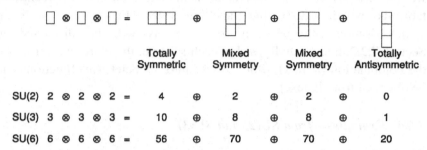

			Totally Symmetric		Mixed Symmetry		Mixed Symmetry		Totally Antisymmetric
SU(2)	2 ⊗ 2 ⊗ 2 =		4	⊕	2	⊕	2	⊕	0
SU(3)	3 ⊗ 3 ⊗ 3 =		10	⊕	8	⊕	8	⊕	1
SU(6)	6 ⊗ 6 ⊗ 6 =		56	⊕	70	⊕	70	⊕	20

Fig. 6.25 Combinations of 3 quarks.

Quark and Antiquark:

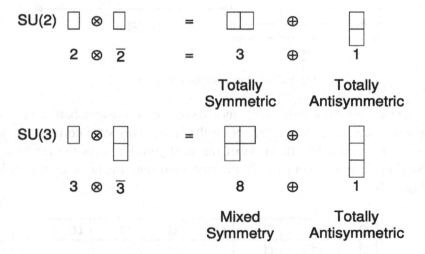

Fig. 6.26 Combinations of a quark and antiquark.

In Section 6.6.1, we saw how the wavefunctions for the distinct states in a multiplet can be identified by inserting single-particle state labels into the boxes. Some care is required for SU(n) with $n > 2$, since for a given tableau, with a given set of labels, there may actually be more than one way of forming wavefunctions with different symmetries – this is related to the observation of multiple nodes at a point in the weight diagrams. For example, with 3 quarks under SU(3)$_{flavour}$, from Fig. 6.25, we find two **8**'s, yet if one inserts labels in boxes, one can only identify 7 tableaux which satisfy the labelling rules. Firstly two different symmetrisation schemes can be used, as will be discussed in Section 7.2.2, and secondly within each scheme, there are two forms for the combination (u, d, s). More insight into the octet wavefunctions can be obtained from [Close].

6.6.4 *Examples – from SU(2) and SU(3)*

For SU(2), the most general tableau has the form as shown in Fig. 6.27.

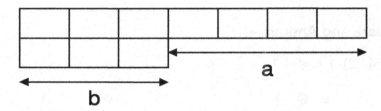

Fig. 6.27 The most general tableau in SU(2).

There cannot be a third row, since there are only two labels, and in a given column, the labels cannot be the same. This state corresponds to $N = a + 2b$ particles. If we label the configuration with the labels for SU(2) of quark flavour (u, d), the first such state might be as shown in Fig. 6.28.

u	u	u	u	u	u	u
d	d	d				

Fig. 6.28 The first state in SU(2).

Since the corresponding wavefunction must be antisymmetrised with respect to the labels in the columns, these will consist of pairs $\sim (ud-du)$.

These correspond to states $I = 0$, $I_3 = 0$. The only interesting part of the tableaux is the part which "sticks out", with just u's in it. Therefore this tableau corresponds to a state with $I = \frac{1}{2} a$, $I_3 = \frac{1}{2} a$. The next state to construct would be as shown in Fig. 6.29.

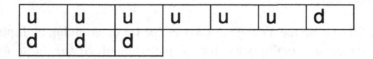

Fig. 6.29 The second state in SU(2).

Effectively, this can be obtained be obtained by applying a lowering operator. The last state in the series will be as shown in Fig. 6.30.

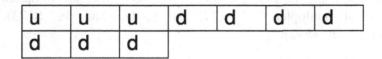

Fig. 6.30 The last state in SU(2).

So there are a total of $(a+1)$ states corresponding to $I = \frac{1}{2} a$. So the multiplicity in SU(2) is:

$$a + 1 \qquad (6.77)$$

For SU(3), the most general tableau has the form as shown in Fig. 6.31.

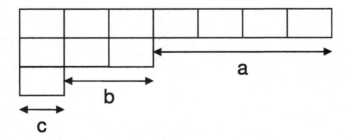

Fig. 6.31 The most general tableau in SU(3).

The multiplicity is:

$$\tfrac{1}{2}(a+1)(b+1)(a+b+2) \tag{6.78}$$

6.7 Problems

Prob. 6.1* Using the Young tableaux recipe for determining multiplicity, write down the multiplicity for N particles of SU(n) in a totally symmetric state, namely a row of N boxes.

By considering examples of how the corresponding states could be labelled by supplying quantum numbers $\{1, 2, \ldots, n\}$ in the N boxes, confirm the expression for the multiplicity.

Prob. 6.2* Using the Young tableaux rules, verify that the multiplicity of a general multiplet in SU(2) is $(a+1)$; and in SU(3), is $\tfrac{1}{2}(a+1)(b+1)(a+b+2)$.

Chapter 7

Hadron States

7.1 Introduction

In this chapter, we will consider the hadron states, as suggested by SU(n): firstly SU(3)$_\text{flavour}$, and then SU(6)$_\text{flavour}\otimes\text{spin}$.

The key message to extract from this chapter is:

- SU(6)$_\text{flavour}\otimes\text{spin}$ provides a description of the low-mass hadron states.

7.2 Hadron States in SU(3)$_\text{flavour}$

Table 7.1 Quantum numbers of the three lightest quarks.

Quark	I	I_3	S	B	$Y =$ $S+B$	$Q =$ $I_3+Y/2$
u	$^1/_2$	$+^1/_2$	0	$^1/_3$	$+^1/_3$	$+^2/_3$
d	$^1/_2$	$-^1/_2$	0	$^1/_3$	$+^1/_3$	$-^1/_3$
s	0	0	-1	$^1/_3$	$-^2/_3$	$-^1/_3$

Table 7.1 and Fig. 7.1 indicate the quantum numbers of the three lightest quarks: u, d and s. Historically, these objects were conceived to explain the observed spectra of light hadrons. The quarks were originally just carriers of quantum numbers, and it was only later that evidence emerged from scattering experiments that they actually corresponded to real particles.

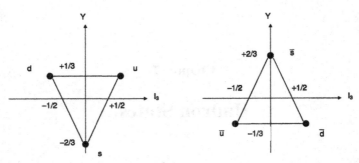

Fig. 7.1 Hypercharge and isospin (3rd component) of the three lightest quarks and antiquarks.

7.2.1 *Mesons*

In the previous chapter, we saw how states could be constructed in $SU(3)_{flavour}$ from a quark (**3**) and an antiquark ($\overline{\textbf{3}}$) resulting in 9 states: an **octet (8)** and a **singlet (1)**. The quark content of these states is shown in Fig. 7.2.

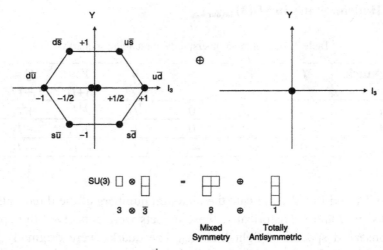

Fig. 7.2 Hypercharge and isospin (3rd component) of the 9 states formed from a quark and an antiquark. The corresponding Young tableaux are also shown.

These 9 states can be associated with 9 of the low-mass mesons, as show in Fig. 7.3, with the flavour wavefunctions shown in Table 7.2.

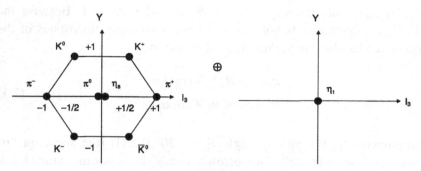

Fig. 7.3 Hypercharge and isospin (3rd component) of the low-mass mesons ($J = 0$).

Table 7.2 Flavour wavefunctions for the low-mass mesons ($J = 0$). S is the hadron strangeness.

	$S = -1$	$S = 0$	$S = +1$
Octet	$K^- = s\bar{u}$	$\pi^- = -d\bar{u}$	
	$\bar{K}^0 = s\bar{d}$	$\pi^0 = \frac{1}{\sqrt{2}}(d\bar{d} - u\bar{u})$	$K^0 = d\bar{s}$
		$\pi^+ = u\bar{d}$	$K^+ = u\bar{s}$
		$\eta_8 = \frac{1}{\sqrt{6}}(u\bar{u} + d\bar{d} - 2s\bar{s})$	
Singlet		$\eta_1 = \frac{1}{\sqrt{3}}(u\bar{u} + d\bar{d} + s\bar{s})$	

The octet states are comparable with the $SU(3)_{\text{colour}}$ gluon states, while the singlet is comparable with the colour description of mesons.

The two η mesons have the same quantum numbers: $I = 0$, $I_3 = 0$, $S = 0$, $B = 0$, $Y = 0$, $Q = 0$ albeit that they have different forms under

SU(3)$_{\text{flavour}}$ – one transforms as an **8**, the other as a **1**. Because the SU(3)$_{\text{flavour}}$ symmetry is not exact, the observed mass eigenstates of the complete Hamiltonian are mixtures of η_8 and η_1:

$$\eta = \cos\theta_p \eta_8 - \sin\theta_p \eta_1$$
$$\eta' = \sin\theta_p \eta_8 + \cos\theta_p \eta_1 \tag{7.1}$$

Experimentally, the mixing angle $\theta_p \approx -20°$ [PDG] (the subscript "p" indicates "pseudoscalar" – see below). The π^0 does not mix since it has $I = 1$.

The particles identified so far are the **pseudoscalar mesons** with $J = 0$, corresponding to a spin state:

$$\frac{1}{\sqrt{2}}(\uparrow\downarrow - \downarrow\uparrow) \tag{7.2}$$

There is a set of heavier particles, the so-called **vector mesons** with $J = 1$. These are indicated in Fig. 7.4.

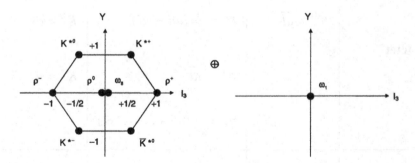

Fig. 7.4 Hypercharge and isospin (3rd component) of the low-mass mesons ($J = 1$).

As with the η mesons, the ω mesons mix:

$$\omega = \cos\theta_v \omega_8 - \sin\theta_v \omega_1$$
$$\varphi = \sin\theta_v \omega_8 + \cos\theta_v \omega_1 \tag{7.3}$$

The mixing is such that ϕ is almost pure $s\bar{s}$:

$$\omega \approx \frac{1}{\sqrt{2}}(u\bar{u} + d\bar{d})$$
$$\phi \approx s\bar{s}$$

(7.4)

As a final remark, historically it was found to be helpful to express the meson wavefunctions in a symmetrised form. Despite there being a quark and an antiquark, these two particles can be treated as different states of a fermion, and therefore the combined wavefunction for the two fermions needs to be antisymmetrised. The flavour symmetry is described by the **G-parity**. For example:

$$\pi^+ = \frac{1}{\sqrt{2}}(u\bar{d} + \bar{d}u) \cdot \frac{1}{\sqrt{2}}(\uparrow\downarrow - \downarrow\uparrow), \quad G = -1$$

(7.5)

$$\rho^+ = \frac{1}{\sqrt{2}}(u\bar{d} - \bar{d}u) \cdot \left\{ \begin{matrix} \uparrow\uparrow \\ \frac{1}{\sqrt{2}}(\uparrow\downarrow + \downarrow\uparrow) \\ \downarrow\downarrow \end{matrix} \right\}, \quad G = +1$$

(7.6)

7.2.2 *Baryons*

The combination of 3 quarks under SU(3) is shown in Fig. 7.5.

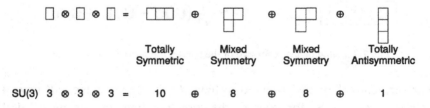

SU(3) 3 ⊗ 3 ⊗ 3 = 10 ⊕ 8 ⊕ 8 ⊕ 1

Fig. 7.5 Young tableaux for the combination of 3 quarks under SU(3).

Can we associate these states with the observed low-mass baryons? A decuplet (**10**), shown in Fig. 7.6, and one octet (**8**), shown in Fig. 7.9, are observed, but there is no obvious singlet and no second octet.

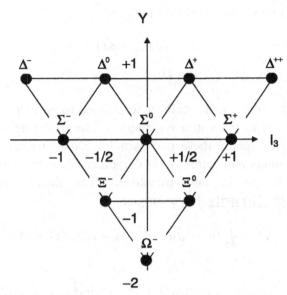

Fig. 7.6 Hypercharge and isospin (3rd component) of the baryon decuplet ($J = {}^3/_2$). The Σ and Ξ states are the heavier ones, i.e. $\Sigma(1385)$ and $\Xi(1530)$.

The decuplet states have a symmetric flavour description illustrated in Fig. 7.7.

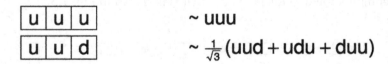

Fig. 7.7 Young tableaux for two of the baryon flavour-symmetric states.

Further, since they are observed to have $J = {}^3/_2$, some states will have a spin state $J_3 = {}^3/_2$. So for example the corresponding state of the Δ^{++} will be $uuu\cdot{\uparrow\uparrow\uparrow}$ – which is a symmetric combination of fermions. The fermionic symmetry is restored by adding the antisymmetric colour wave-function:

$$\tfrac{1}{\sqrt{6}}(rbg + grb + bgr - rgb - brg - gbr) \qquad (7.7)$$

The complete list of $J = {}^3/_2$ baryons is given in Fig. 7.8.

Quark content	Baryon	I	I_3	S
u u u	Δ^{++}	3/2	+3/2	0
u u d	Δ^{+}	3/2	+1/2	0
u d d	Δ^{0}	3/2	−1/2	0
d d d	Δ^{-}	3/2	−3/2	0
u u s	Σ^{+}	1	+1	−1
u d s	Σ^{0}	1	0	−1
d d s	Σ^{-}	1	−1	−1
u s s	Ξ^{0}	1/2	+1/2	−2
d s s	Ξ^{-}	1/2	−1/2	−2
s s s	Ω^{-}	0	0	−3

Fig. 7.8 The $J = {}^3/_2$ baryon decuplet. The Σ and Ξ states are the heavier ones, i.e. Σ (1385) and Ξ (1530).

The baryon octet is shown in Fig. 7.9. These states have $J = {}^1/_2$.

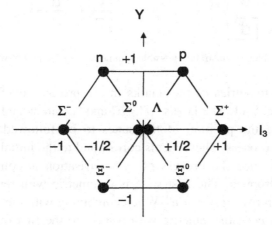

Fig. 7.9 Hypercharge and isospin (3rd component) of the baryon octet ($J = {}^1/_2$). The Σ and Ξ states are the lighter ones.

In trying to understand the origin of and wavefunctions for the octet, we recall how a third quark is added to the state formed by a pair of quarks – this is illustrated in Fig. 7.10.

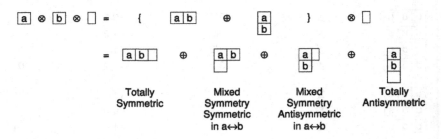

Totally Symmetric   Mixed Symmetry Symmetric in a↔b   Mixed Symmetry Antisymmetric in a↔b   Totally Antisymmetric

Fig. 7.10 Young tableaux depicting the addition of a third quark to states formed by an existing pair of quarks. The labels "*a*" and "*b*" depict the first and second quark states.

So we see the two octets are formed with different symmetry for the first two quarks. So if we consider a proton, which has a flavour content *uud*, then although the flavour labels can only be inserted into the octet tableau in one way, the way in which we symmetrise the wavefunction will depend on which 2 quarks we consider as the first pair.

Fig. 7.11 Young tableaux illustrating possible symmetries for the proton wavefunction.

The possible symmetries of the quarks in the proton wavefunction are illustrated in Fig. 7.11. The labels "*a*", "*b*" and "*c*" are added to the state labels to enable the position of the states to be followed during the symmetrisation process. These labels correspond to the initial position of the quarks, but once the process of symmetrisation is completed, the letters can be dropped. The first state is symmetric with respect to the quarks in the top row: $u^a u^b + u^b u^a$. When combining with the third quark, the state should be antisymmetric with respect to the first quarks in the wavefunction – those quarks in the left-hand column (and remember, as result of the symmetrisation, the two *u*-quarks swap places):

$$p \sim (u^a u^b + u^b u^a) \otimes d^c$$
$$= u^a u^b \otimes d^c + u^b u^a \otimes d^c$$
$$= (u^a u^b d^c - d^c u^b u^a) + (u^b u^a d^c - d^c u^a u^b) \qquad (7.8)$$
$$\sim uud - duu$$

The $\otimes$ is used to indicate a combination involving multiplication as well as the appropriate symmetrisation.

The second state is antisymmetric with respect to the quarks in the left-hand column: $u^a d^b - d^b u^a$. When combining with the third quark, the state should be symmetric with respect to those quarks in the top row:

$$p \sim (u^a d^b - d^b u^a) \otimes u^c$$
$$= u^a d^b \otimes u^c - d^b u^a \otimes u^c$$
$$= (u^a d^b u^c + u^c d^b u^a) - (d^b u^a u^c + u^c u^a d^b) \qquad (7.9)$$
$$\sim 2udu - duu - uud$$

A slightly different way of formulating this is to consider that the boxes in the Young tableau in Fig. 7.11 relate to the quarks – and to these we attach labels. In the first step, we antisymmetrise with respect to the 1st quark (u^a) and 3rd quark (d^b), leaving a place-holder "*" for the 2nd quark. Then we symmetrise with respect to the 1st quark (u^a or d^b) and 2nd quark (u^c), replacing the "*":

$$p \sim (u^a * d^b - d^b * u^a) \otimes u^c$$
$$= u^a * d^b \otimes u^c - d^b * u^a \otimes u^c$$
$$= (u^a u^c d^b + u^c u^a d^b) - (d^b u^c u^a + u^c d^b u^a) \qquad (7.10)$$
$$\sim 2uud - duu - udu$$

The wavefunction in Eq. (7.8), is antisymmetric in $1 \leftrightarrow 3$; Eq. (7.9) is symmetric in $1 \leftrightarrow 3$. Usually the symmetry is expressed with respect to the first two particles, and so we recast Eqs (7.8) and (7.9):

$$\frac{1}{\sqrt{2}}(ud - du)u$$
$$\text{and} \qquad (7.11)$$
$$\frac{1}{\sqrt{6}}\{(ud + du)u - 2uud\}$$

Symmetries and Conservation Laws in Particle Physics

So which is the wavefunction for a proton? There appears to be an ambiguity, and this will be resolved when we look at SU(6). For the octet states with two identical quark flavours, we will find analogous wavefunctions to the above (with the same ambiguity). The observed baryon states have $J = {}^1/_2$ and are shown in Fig. 7.12.

Quark content	Baryon	I	I_3	S
u u / d	p	1/2	+1/2	0
u d / d	n	1/2	−1/2	0
u u / s	Σ^+	1	+1	−1
u d / s	Σ^0	1	0	−1
d d / s	Σ^-	1	−1	−1
u d / s	Λ	0	0	−1
u s / s	Ξ^0	1/2	+1/2	−2
d s / s	Ξ^-	1/2	−1/2	−2

Fig. 7.12 The $J = {}^1/_2$ baryon octet.

7.3 Hadron States in SU(6)$_{\text{flavour}\otimes\text{spin}}$

As we have just seen for the baryons, in SU(3)$_{\text{flavour}}$, it is not obvious
  a. how to assign the octet wavefunctions
  b. why there is no flavour singlet

Rather than separately combining flavour and spin: SU(3)$_{\text{flavour}}$ $\otimes$ SU(2)$_{\text{spin}}$, consisting of $\{u, d, s\} \otimes \{\uparrow, \downarrow\}$, we consider SU(6)$_{\text{flavour}\otimes\text{spin}}$, consisting of the states $\{u\uparrow, d\uparrow, s\uparrow, u\downarrow, d\downarrow, s\downarrow\}$ – all of which are considered indistinguishable. As before, the flavour-spin states will be combined with an SU(3)$_{\text{colour}}$ singlet. Since the latter is antisymmetric, the flavour$\otimes$spin states will need to be symmetric if they are to describe identical fermions. Although it is debatable as to whether quarks with different flavours can be considered "identical", undeniably this must be addressed for the baryons with a flavour content *uuu*, *ddd* or *sss*. With 3 quarks having 6 possible states, there are $6^3 = 216$ possible states – the Young tableaux will be identical to those of Fig. 7.5, however the multiplicities will be 56, 70, 70 and 20 (easily checked using the recipe in Section 6.6.3). The **56** will be totally symmetric and hence a suitable candidate for the baryon states.

The wavefunctions for the **56** are obtained from appealing to the states derived from $SU(3)_{flavour} \otimes SU(2)_{spin}$, shown in Fig. 7.13. The combinations of these states lead to new symmetries, as shown Table 7.3.

$SU(3)_{flavour}$	$3 \otimes 3 \otimes 3 =$	10	$\oplus$	8	$\oplus$	8	$\oplus$	1
		ϕ_S		$\phi_{M,S}$		$\phi_{M,A}$		ϕ_A
$SU(2)_{spin}$	$2 \otimes 2 \otimes 2 =$	4	$\oplus$	2	$\oplus$	2		
		χ_S		$\chi_{M,S}$		$\chi_{M,A}$		

Fig. 7.13 Classes of wavefunctions obtained from $SU(3)_{flavour} \otimes SU(2)_{spin}$ for 3 quarks. The flavour states are denoted by ϕ's; the spin states by χ's. ϕ_S is symmetric under exchange of all 3 quarks; $\phi_{M,S}$ has mixed symmetry and is symmetric under exchange of only 2 quarks, etc..

Table 7.3 Resultant symmetries of wavefunctions arising from the combinations of flavour (ϕ) and spin (χ) wavefunctions of particular symmetry. "S", "M", "A" denote symmetric, mixed and antisymmetric states respectively.

		χ_{spin}	
		S	M
$\phi_{flavour}$	S	S	M
	M	M	S, M, A
	A	A	M

We can see from Table 7.3, there are two sets of combinations which result in flavour-spin wavefunctions which are <u>totally</u> symmetric. This is obvious for the combination of totally symmetric states (ϕ_S, χ_S), although not at all obvious for the states of mixed symmetry. The resulting wavefunctions are:

$$\phi_S \chi_S$$
$$\frac{1}{\sqrt{2}}(\phi_{M,S}\chi_{M,S} + \phi_{M,A}\chi_{M,A}) \tag{7.12}$$

The first set of wavefunctions corresponds to 10 flavour states combined with 4 spin states, giving 40 states; the second set corresponds to 8 flavour states combined with 2 spin states, giving 16 states – a total of 56 states, corresponding to the **56** of SU(6). It is instructive to construct some of these states and verify their symmetry.

Decuplet states:

$$\Delta^{++}, J_3 = \tfrac{3}{2}: \quad uuu \uparrow\uparrow\uparrow \tag{7.13}$$

$$\Delta^+, J_3 = \tfrac{1}{2}: \quad \tfrac{1}{\sqrt{3}}(uud + udu + duu)\tfrac{1}{\sqrt{3}}(\uparrow\uparrow\downarrow + \uparrow\downarrow\uparrow + \downarrow\uparrow\uparrow) \tag{7.14}$$

Octet states:

$$p, J_3 = \tfrac{1}{2}:$$

$$\tfrac{1}{\sqrt{2}} \cdot \tfrac{1}{\sqrt{6}}\{(ud + du)u - 2uud\} \cdot \tfrac{1}{\sqrt{6}}\{(\uparrow\downarrow + \downarrow\uparrow)\uparrow - 2\uparrow\uparrow\downarrow\} +$$

$$\tfrac{1}{\sqrt{2}} \cdot \tfrac{1}{\sqrt{2}}(ud - du)u \cdot \tfrac{1}{\sqrt{2}}(\uparrow\downarrow - \downarrow\uparrow)\uparrow$$

$$= \tfrac{1}{\sqrt{2}}\{+\tfrac{2}{3}udu - \tfrac{1}{3}duu - \tfrac{1}{3}uud\}\uparrow\downarrow\uparrow +$$

$$\tfrac{1}{\sqrt{2}}\{-\tfrac{1}{3}udu + \tfrac{2}{3}duu - \tfrac{1}{3}uud\}\downarrow\uparrow\uparrow +$$

$$\tfrac{1}{\sqrt{2}}\{-\tfrac{1}{3}udu - \tfrac{1}{3}duu + \tfrac{2}{3}uud\}\uparrow\uparrow\downarrow$$

$$= \tfrac{1}{\sqrt{2}}\{u\uparrow u\uparrow d\downarrow + u\uparrow d\downarrow u\uparrow + d\downarrow u\uparrow u\uparrow$$

$$- \tfrac{1}{3}(uud + udu + duu)(\uparrow\uparrow\downarrow + \uparrow\downarrow\uparrow + \downarrow\uparrow\uparrow)\} \tag{7.15}$$

This is pretty messy, but obviously symmetric. A complete list of the flavour wavefunctions can be found in [Close].

One can play the same games with the mesons under SU(6):

$$\mathbf{6} \otimes \overline{\mathbf{6}} = \mathbf{35} \oplus \mathbf{1} \qquad (7.16)$$

However, this formulation seems less helpful, in particular as we seem to have an appropriate description of the low-mass mesons already.

7.4 Some Final Comments on Hadrons

In addition to the light quarks, $\{u, d, s\}$, there are 3 **heavier quarks**. In principle, one can extend the symmetry groups to include $\{c, b, t\}$. However, since their masses are so much heavier and their masses exceed the QCD scale, Λ_{QCD}, the symmetry is broken badly. Instead, it is better to treat the heavy quarks, $\{Q\}$, separately from the lighter ones $\{q\}$. Heavy-flavour mesons can be described by $Q\overline{q}$ where q transforms according to SU(3); heavy baryons by $Qq^a q^b$ where q^a and q^b transform according to SU(3).

So far, all the hadron states considered have orbital angular momentum $L = 0$ (s-wave). Excited states can be obtained by considering higher orbital angular momentum.

7.5 Problems

Prob. 7.1 The proton wavefunction ($J_3 = +^1/_2$) in terms of flavour and spin is:

$$\frac{1}{\sqrt{2}}\frac{1}{\sqrt{6}}\{(ud+du)u-2uud\}\frac{1}{\sqrt{6}}\{(\uparrow\downarrow+\downarrow\uparrow)\uparrow-2\uparrow\uparrow\downarrow\}+$$
$$\frac{1}{\sqrt{2}}\frac{1}{\sqrt{2}}(ud-du)u\frac{1}{\sqrt{2}}(\uparrow\downarrow-\downarrow\uparrow)\uparrow$$

Apply the isospin lowering operator $I_- = I_-^a + I_-^b + I_-^c$ (superscripts refer to 1st, 2nd and 3rd quarks) to this wavefunction to obtain the neutron description, expressing its wavefunction in a form to make the similarity manifest.

Prob. 7.2* The magnetic moment operator for a single quark is proportional to the product of the charge and the spin operators: QJ.

When considering a baryon, since the wavefunction is symmetrised with respect to all three quarks, it will suffice to consider only the third quark (and multiply the result by 3). Find the ratios of the magnetic moments of the proton and neutron, by considering the matrix elements for the operator $Q^c \sigma_3^c$ (the superscript "c" indicates that these operators only act on the third quark). Use the wavefunctions from Prob. 7.1. The measured value of this ratio is -1.46.

Prob. 7.3* An alternative approach to calculating baryon magnetic moments is to use simplified wave-functions along with the magnetic moment operators for all three quarks [Lichtenberg].

Using:

$$p = uud \; \frac{1}{\sqrt{6}} ((\uparrow\downarrow + \downarrow\uparrow) \uparrow - 2 \uparrow\uparrow\downarrow)$$

and a magnetic moment operator:

$$M = Q^a J_3^a + Q^b J_3^b + Q^c J_3^c$$

where Q^i is the charge operator for the ith quark and J_3^i is its 3rd component of spin, calculate an expression for the proton's magnetic moment $\langle p|M|p \rangle$. (Use q_u, q_d, q_s for the quark charges – do not express them as numbers.)

Use isospin symmetry to calculate the neutron's magnetic moment. Finally, using:

$$\Lambda = uds \; \frac{1}{\sqrt{2}} (\uparrow\downarrow - \downarrow\uparrow) \uparrow$$

calculate the magnetic moment of the Λ baryon.

Prob. 7.4 In the Hamiltonian for the meson masses, there is a term $\kappa J^a \cdot J^b$, where $J^{a,b}$ are the quark spins. Calculate the value of the mass splitting between the vector and pseudoscalar mesons induced by this term.

Prob. 7.5 Using Young tableaux, verify the multiplicities in Eq. (7.16).

Chapter 8

The Standard Model and Beyond

8.1 Introduction

In this chapter, we will consider some of the **consequences** of group theory and see how it plays a part in the **Standard Model** of Particle Physics and models **beyond the Standard Model**.

The key messages to extract from this chapter are:

- Group theory, although apparently rather esoteric, helps us understand key areas of Particle Physics.
- The concepts of group theory underpin the Standard Model, in particular the attribution of quantum numbers.
- Group theory helps motivate models beyond the Standard Model.

8.2 Consequences of Group Theory

We have already seen how group theory impacts on

- **angular momentum**
- the description of **hadronic states**
- the description of the **gauge forces** and **scattering amplitudes**

There are many other consequences which will not be investigated in this book. In particular, the description of the hadronic states allows many useful calculations to be made, albeit that these were of greater interest in the past, and less of a focus in the LHC-era:

i) Assuming all the light quarks really are indistinguishable, then the masses of members of the multiplets should be identical, in the approximation that strong interactions dominate the interactions between the constituent quarks. Allowing for differences in the bare quark masses and their electromagnetic interactions, improved estimates of the relationships between the **hadron masses** can be made, for example the Gell-Mann Okubu formula [Lichtenberg].

ii) Relative **magnetic moments** can be calculated – see Probs 7.2 & 7.3.

iii) Relative **decay rates** can be calculated.

8.3 The Standard Model

The starting point for the Standard Model is the Lagrangian for massless fermions:

$$L \sim \overline{\psi}\gamma_{\mu}\partial^{\mu}\psi \tag{8.1}$$

On this, we impose invariance under local gauge transformations associated with a "charge" g. As explained in Sections 3.2.3 and 4.5, this is achieved through the introduction of the covariant derivative:

$$\partial^{\mu} \rightarrow D^{\mu} = \partial^{\mu} - igX \cdot W^{\mu} \tag{8.2}$$

where X is the generator of the gauge transformation and W is the field required to maintain the invariance. As there may be more than one generator, there is an implicit scalar product (summation) over the different degrees of freedom. The Standard Model is constructed from the gauge groups:

$$\text{U(1)}_{\text{weak hypercharge}} \otimes \text{SU(2)}_{\text{weak isospin}} \otimes \text{SU(3)}_{\text{colour}} \tag{8.3}$$

Table 8.1 lists the three gauge forces. The generators of SU(2) and SU(3) are the Pauli and Gell-Mann matrices, respectively, whilst that of U(1) is simply a number equal to the particle hypercharge.

Table 8.1 The three gauge forces.

"Charge"	Symmetry Group	Generator	Coupling	Boson Fields
Weak hypercharge	U(1)	Y	g_Y	B
Weak isospin	SU(2)	T	g_T	W^+, W^0, W^-
Quark colour	SU(3)	λ	g_c	$G_{1, 2, \ldots, 8}$

The fermion content of the SM Lagrangian is given by a set of **left-handed** lepton and quark **doublets** under SU(2)$_\text{weak isospin}$:

$$L_L = \left\{ \begin{pmatrix} \nu_e \\ e^- \end{pmatrix}_L, \begin{pmatrix} \nu_\mu \\ \mu^- \end{pmatrix}_L, \begin{pmatrix} \nu_\tau \\ \tau^- \end{pmatrix}_L \right\}$$

$$Q_L = \left\{ \begin{pmatrix} u \\ d \end{pmatrix}_L, \begin{pmatrix} c \\ s \end{pmatrix}_L, \begin{pmatrix} t \\ b \end{pmatrix}_L \right\} \tag{8.4}$$

and a set of **right-handed** lepton and quark **singlets**:

$$\nu_R = \{ \nu_{eR}, \nu_{\mu R}, \nu_{\tau R} \}$$
$$l_R = \{ e_R^-, \mu_R^-, \tau_R^- \}$$
$$u_R = \{ u_R, c_R, t_R \} \tag{8.5}$$
$$d_R = \{ d_R, s_R, b_R \}$$

Furthermore:
- Each quark is a colour triplet: $q = (q_r, q_b, q_g)$.
- All particle states are Dirac spinors (4 components).

So all of the particles are complicated tensors, and implicitly carry several indices.

Following the observations of **neutrino oscillations**, the neutrinos must be massive and there must exist right-handed states. This is claimed to be evidence for **New Physics beyond the Standard Model**. However, the Standard Model (which originally was constructed to include a description of massless neutrinos) can trivially be extended to include right-handed states and masses. Having said this, because of a hierarchy problem associated with the neutrino masses being much less than those of the charged leptons and quarks, many theorists prefer the idea of neutrinos with a **Majorana** mass term in the Lagrangian (for which the neutrino is its own antiparticle). The masses of the neutrinos are then manifestations of new physics at the GUT or Planck scale. Nevertheless, we will stick with the Dirac description of neutrinos for simplicity.

The quantum numbers of the fermions are indicated in Table 8.2.

Table 8.2 Fermion quantum numbers. The charge is given by $Q = T_3 + \frac{1}{2}Y$.

	Hypercharge Y	Isospin T	Colour λ
L_L	-1	$\frac{1}{2}$	0
Q_L	$\frac{1}{3}$	$\frac{1}{2}$	1
ν_R	0	0	0
l_R	-2	0	0
u_R	$\frac{4}{3}$	0	1
d_R	$-\frac{2}{3}$	0	1

So the Standard Model Lagrangian is:

$$
\begin{aligned}
L = \quad & \overline{L}_L && \gamma_\mu(\partial^\mu && -ig_Y B^\mu && +ig_T T\cdot W^\mu) && && L_L \\
+ \quad & \overline{Q}_L && \gamma_\mu(\partial^\mu && +\tfrac{1}{3}ig_Y B^\mu && +ig_T T\cdot W^\mu && +\tfrac{1}{2}ig_c \lambda\cdot G^\mu) && Q_L \\
+ \quad & \overline{v}_R && \gamma_\mu(\partial^\mu) && && && && v_R \\
+ \quad & \overline{l}_R && \gamma_\mu(\partial^\mu && -2ig_Y B^\mu) && && && l_R \\
+ \quad & \overline{u}_R && \gamma_\mu(\partial^\mu && +\tfrac{4}{3}ig_Y B^\mu && && +\tfrac{1}{2}ig_c \lambda\cdot G^\mu) && u_R \\
+ \quad & \overline{d}_R && \gamma_\mu(\partial^\mu && -\tfrac{2}{3}ig_Y B^\mu && && +\tfrac{1}{2}ig_c \lambda\cdot G^\mu) && d_R
\end{aligned}
$$

+ boson terms
+ Higgs terms, rendered Gauge Invariant, giving Boson mass terms
+ Higgs–Fermion terms, giving Fermion mass terms

$$(8.6)$$

where the Y quantum numbers have been explicitly inserted in the B^μ terms; ½ is required in associated with the Gell-Mann matrices; and the isospin generators T implicitly include a factor of ½ in association with the Pauli spin matrices.

Baryon and lepton number conservation is explicitly built into the model. We see that the v_R has no gauge couplings – it is "sterile" – the only interactions it has are with the Higgs.

Could $U(1)_Y = U(1)_{EM}$? No – because terms like $\overline{v}_L B v_L$ indicate a coupling of the neutrino to the $U(1)_Y$ field B, and since the neutrino has no electrical charge, it does not couple to the exchange boson of $U(1)_{EM}$, namely the photon.

8.3.1 *Quantum numbers*

In order to be gauge invariant, that is: unaffected by the symmetry transformations corresponding to $U(1)_Y \otimes SU(2)_T \otimes SU(3)_c$, terms in the Lagrangian must carry no net quantum numbers. So we can deduce that the B boson corresponding to $U(1)_Y$ must have $Y = 0$ and $T = 0$, while the W bosons have $Y = 0$, $T = 1$ and $Q = +1, 0$ or -1.

So for example, consider a term like:

$$\bar{e} W^- \nu_e \qquad (8.7)$$

corresponding to a vertex illustrated in Fig. 8.1.

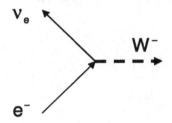

Fig. 8.1 Vertex for a W boson coupling to an electron and neutrino.

The weak isospin must be conserved:

$$
\begin{array}{ccccc}
T_3(e^-) & \rightarrow & T_3(\nu_e) & + & T_3(W^-) \\
-{}^1\!/_2 & \rightarrow & {}^1\!/_2 & + & -1
\end{array}
$$

8.4 The Higgs Mechanism

The Higgs mechanism is well explained in many text books. Here, we will focus on the description which arises in the context of $SU(2)_T$ and the corresponding quantum numbers.
The Higgs doublet:

$$\begin{pmatrix} \phi^+ \\ \phi^0 \end{pmatrix} \qquad (8.8)$$

has $Y = 1$, $T = {}^1\!/_2$ – this choice ensures $Q = T_3 + \frac{1}{2}Y$. We recall for SU(2), that the **2** transforms like the **2**. So the conjugate state is:

$$\begin{pmatrix} \bar{\phi}^0 \\ -\phi^- \end{pmatrix} \qquad (8.9)$$

where the components are the complex conjugates of the components of the **2**. The **2** has $Y = -1$, $T = {}^1\!/_2$.

In the Higgs Mechanism, the real part of ϕ^0 is written as $v+h$, where v (an italicised "v", not to be confused with a Greek nu, ν) is the Higgs vacuum expectation value (vev) and h is the real Higgs field. ϕ^+ and ϕ^- and the imaginary part of ϕ^0 have vanishing vev's and the fields correspond to the Goldstone bosons, which are absorbed by the W bosons to provide their longitudinal polarisation. This gives:

$$\mathbf{2} \sim \begin{pmatrix} 0 \\ v+h \end{pmatrix} \quad \text{and} \quad \overline{\mathbf{2}} \sim \begin{pmatrix} v+h \\ 0 \end{pmatrix} \qquad (8.10)$$

The Higgs Mechanism results from imposing $U(1) \otimes SU(2)$ gauge invariance on the Higgs Lagrangian terms:

$$L_{\text{Higgs}} \sim (\partial_\mu \phi^H)(\partial^\mu \phi) + V(\phi) \qquad (8.11)$$

This is achieved simply by replacing the derivative by the covariant derivative D^μ. Expanding this expression and looking just at the vacuum expectation gives terms like $v^2 WW$ – these look like W mass terms in a Lagrangian:

$$L_{\text{W mass}} \sim m_W^{\ 2} WW \qquad (8.12)$$

By construction, the Higgs mechanism provides masses for the vector bosons $\{W^+, W^0, W^-\}$ and causes the mixing of the W^0 and B gauge fields, so producing the Z and γ fields.

The part of the Lagrangian describing the fermion masses (ignoring the different couplings which lead to different fermion masses) is:

$$
\begin{aligned}
L_{\text{fermion mass}} \sim\ & \overline{L}_L \begin{pmatrix} \phi^+ \\ \phi^0 \end{pmatrix} l_R + \overline{L}_L \begin{pmatrix} \overline{\phi}^0 \\ -\phi^- \end{pmatrix} \nu_R + \\
& \overline{Q}_L \begin{pmatrix} \phi^+ \\ \phi^0 \end{pmatrix} d_R + \overline{Q}_L \begin{pmatrix} \overline{\phi}^0 \\ -\phi^- \end{pmatrix} u_R + hc +
\end{aligned}
\qquad (8.13)
$$

non-diagonal terms

The above expression needs to be repeated for each of the generations (ν_e, e), (ν_μ, μ), (ν_τ, τ), (u, d), (c, s), (t, b). "Non-diagonal" terms which combine different generations give rise to mixing, as found in the CKM matrix. The fermion-mass terms appear when the ϕ fields are replaced by their vev's, giving terms like:

$$v\bar{l}_L l_R \tag{8.14}$$

which looks like the lepton mass term:

$$L_{\text{lepton mass}} \sim m_l \bar{l}_L l_R \tag{8.15}$$

The second term in Eq. (8.13) is:

$$\bar{v}_L \bar{\phi}^0 v_R - \bar{l}_L \phi^- v_R \tag{8.16}$$

This is new for the Standard Model, and the first part of this corresponds to Dirac masses for the neutrinos (if this is the correct mechanism for neutrino masses).

The different assignment of weak quantum numbers to the left- and right-handed fermions allows the possibility of C, P and CP violation. As CPT is conserved in quantum field theories, this implies T can also be violated.

8.5 Beyond the Standard Model

In this section, we will touch on some of the theories beyond the Standard Model of Particle Physics for completeness. The emphasis will be on the corresponding groups of these models. The interested reader is referred to [Georgi, Barnes] for more details.

While the Standard Model is great, in so far as it describes many features of Particle Physics and provides accurate predictions of many measured quantities, it has several undesirable features:

- a large number of unconstrained **constants** (couplings, charges)
- no explanation of **generations**
- no explanation as to **charge quantisation**:

$$Q(u) = -\tfrac{2}{3}Q(e^-) \text{ and } Q(d) = \tfrac{1}{3}Q(e^-) \tag{8.17}$$

It would seem *elegant* to contain all of the SM in some single theory (group) – rather than as the product of three seemingly disconnected groups. Of course, just because something is elegant does not make it true; the truth needs to be determined experimentally. While the electroweak interaction is supposedly unified, nevertheless it is represented by the product of two groups $SU(2)_T$ and $U(1)_Y$, albeit that they are mixed through the Higgs Mechanism.

If the **coupling constants** of the three groups $U(1)_Y$, $SU(2)_T$ and $SU(3)_c$ are evolved via the Renormalisation Group Equations (RGE), they become equal at a scale of ~10^{15} GeV. It turns out that this convergence is even more precise if **SuperSymmetry** is included. So there are reasons to presume that there is indeed something beyond the Standard Model. SU(5) is one such model.

8.5.1 *SU(5)*

In SU(5), the fundamental representation is taken as:

$$
\mathbf{5} = \begin{pmatrix} d_r \\ d_b \\ d_g \\ e^+ \\ \overline{\nu}_e \end{pmatrix}_R \tag{8.18}
$$

where the subscripts r, b, g refer to colour. The particles of the Standard Model (in the absence of neutrino mass, and hence with no ν_R) can be contained in the multiplets of SU(5):

$$
\overline{\mathbf{5}} = \begin{pmatrix} \overline{d}_r \\ \overline{d}_b \\ \overline{d}_g \\ e^- \\ -\nu_e \end{pmatrix}_L \tag{8.19}
$$

$$10 = \begin{pmatrix} 0 & \bar{u}_g & -\bar{u}_b & -u_r & -d_r \\ -\bar{u}_g & 0 & \bar{u}_r & -u_b & -d_b \\ \bar{u}_b & -\bar{u}_r & 0 & -u_g & -d_g \\ u_r & u_b & u_g & 0 & -e^+ \\ d_r & d_b & d_g & e^+ & 0 \end{pmatrix}_L \qquad (8.20)$$

where this is reproduced for each of the three generations (v_e, e^-, u, d), (v_μ, μ^-, c, s) and (v_τ, τ^-, t, b). The **10** is an antisymmetric multiplet derived from $5 \otimes 5$, as shown in Fig. 8.2.

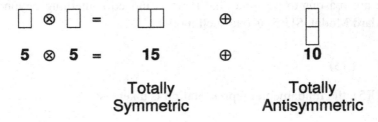

5 ⊗ 5 = 15 ⊕ 10

Totally   Totally
Symmetric   Antisymmetric

Fig. 8.2 Multiplets found in SU(5).

There are $5^2 - 1 = 24$ gauge bosons. In terms of the fundamental representation, they have the form shown in Table 8.3.

Table 8.3 Combination of quantum numbers from the fundamental representation corresponding to the gauge bosons.

	d_r	d_b	d_g	e^+	$\bar{v}_e$
d_r	g, γ, Z	g	g	X_r	Y_r
d_b	g	g, γ, Z	g	X_b	Y_b
d_g	g	g	g, γ, Z	X_g	Y_g
e^+	X_r	X_b	X_g	γ, Z	W^+
$\bar{v}_e$	Y_r	Y_b	Y_g	W^-	Z

So there are 8 gluons, g; 3 weak bosons, W^+, W^-, Z; and 1 photon, γ; along with 12 X, Y bosons. The X, Y bosons are coloured and transform leptons to quarks and v.v. – they are leptoquarks. As we saw in Chapter 3, the X, Y bosons can mediate proton decay, where the quantum numbers B and L are not conserved, but $B - L$ is.

The points in favour of SU(5) are:

- It is the smallest group which contains $U(1)_Y \otimes SU(2)_T \otimes SU(3)_c$, and therefore has the greatest predictive power.
- T_3 and Y are both generators of SU(5) and hence traceless. (In the Standard Model, Y was a generator of $U(1)_Y$, and was not traceless). Since $Q = T_3 + \frac{1}{2}Y$, Q is traceless and in the fundamental representation this leads to:

$$Q(d_r) + Q(d_b) + Q(d_g) + Q(e^+) + Q(\overline{v}_e) = 0$$

$$\Rightarrow 3Q(d) + Q(e^+) = 0 \tag{8.21}$$

$$\Rightarrow Q(d) = \tfrac{1}{3}Q(e^-)$$

- It predicts $\sin^2 \theta_W$ very accurately.
- The couplings of the low-energy gauge group representations converge at a scale of the X boson mass $M(X)$.
- Since there is only room for one neutrino type in the $\overline{\mathbf{5}}$ and $\mathbf{10}$, i.e. there is no v_R, the only way to generate a neutrino mass is via a Majorana term which requires $v = \overline{v}$ – which violates $B - L$ conservation. Hence the neutrino must be massless. This used to be a good feature in the days where measurements suggested the neutrino might indeed be massless.

The problems with SU(5) are:

- The neutrino now appears to have mass.
- The predicted decay rate for the proton is much larger than the current observed limits [PDG].
- There is no explanation of the three generations.

So despite its elegance, SU(5) is now excluded.

8.5.2 *Other symmetry groups*

Another possible group which could contain the Standard Model gauge groups is SO(10) – this contains SU(5) and is not yet excluded. Going further, other possibilities include the **exceptional groups**: E_6 and E_8. The latter is of relevance to the heterotic superstring (combining bosonic and fermionic modes). These groups are associated with the algebra of **octonians**, which are generalisations of the imaginary number $i = \sqrt{-1}$. E_6 contains SO(10) [Georgi].

8.5.3 *SuperSymmetry*

SUSY provides an extension of the Poincaré groups (associated with translations, rotations and Lorentz boosts). It is the one remaining symmetry in quantum field theory which is not exploited in the Standard Model Lagrangian. Under SUSY, transformations take place which treat bosons and fermions as indistinguishable states. Crudely speaking, the generator can be thought of as the square root of the momentum operator! For each standard particle, there is a superpartner, with the same couplings but with a different mass and differing in spin by half a unit. The corresponding super-field doublets are:

$$\text{gauge:} \begin{pmatrix} J = 1 \\ J = \frac{1}{2} \end{pmatrix}, \quad \text{chiral (matter):} \begin{pmatrix} J = \frac{1}{2} \\ J = 0 \end{pmatrix} \qquad (8.22)$$

where the upper component in each doublet is the SM particle and the lower component is the SUSY partner.

In many SUSY models, there is a conserved quantity, **R-parity**, which distinguishes between the standard particles and their super-partners. This ensures that SUSY particles can only be created in pairs, and that once created, a SUSY particle will always leave a SUSY particle in the final state, leading to a stable **lightest supersymmetric particle** (LSP). The LSP is usually considered to be the neutralino – a mixture of the photino, Zino and neutral Higgsinos, the superpartners of the photon, Z boson and neutral Higgs boson, respectively. The LSP may be one of the constituents of **dark matter**.

Appendix

Hints and Answers to Problems

This appendix contains hints and outline answers for some of the problems set in the chapters.
For teachers, a complete set of outline answers can be provided upon request to the author.

Prob. 1.1
You may find it informative to look at the Baker–Campbell–Hausdorf formula, for example on the web.

Prob. 1.2
This can be solved by expanding the exponential. Alternatively, one can diagonalise A: $A = E\Lambda E^{-1}$ and then it is easy to prove:

$$\exp(i\alpha A) = E \exp(i\alpha\Lambda)E^{-1}$$

Since the matrix of eigenvalues, Λ, is diagonal, the exponential is the diagonal matrix of the exponentials of the eigenvalues $(+1, 0, -1)$. So:

$$\exp(i\alpha A) = E\{\exp(i\alpha\lambda_i)\}E^{-1}$$

$$= \frac{1}{\sqrt{2}}\begin{pmatrix} 1 & 0 & 1 \\ 0 & \sqrt{2} & 0 \\ 1 & 0 & -1 \end{pmatrix}\begin{pmatrix} \exp(i\alpha) & 0 & 0 \\ 0 & 1 & 0 \\ 0 & 0 & \exp(-i\alpha) \end{pmatrix}\frac{1}{\sqrt{2}}\begin{pmatrix} 1 & 0 & 1 \\ 0 & \sqrt{2} & 0 \\ 1 & 0 & -1 \end{pmatrix}$$

$$= \begin{pmatrix} \cos\alpha & 0 & i\sin\alpha \\ 0 & 1 & 0 \\ i\sin\alpha & 0 & \cos\alpha \end{pmatrix}$$

Prob. 1.3
For the expression:

$$f(\alpha) = \exp(i\alpha A)B\exp(-i\alpha A)$$

it would be helpful to move the first exponential past B to "annihilate" on the second exponential. If we expand the first exponential, we will have terms like $A^n B$. Using the commutation relationship:

$$[A, B] = B$$
$$\Rightarrow AB = BA + B = B(A + I)$$
$$\Rightarrow A^n B = B(A + I)^n$$

So:

$$\exp(i\alpha A)B = (1 + i\alpha A - \alpha^2 A^2 / 2! - ...)B$$
$$= B(1 + i\alpha(A + I) - \alpha^2 (A + I)^2 / 2! - ...) = B\exp(i\alpha(A + I))$$

The two exponents trivially commute, and hence can be added, so:

$$f(\alpha) = B\exp(i\alpha(A + I))\exp(-i\alpha A) = B\exp(i\alpha)$$

David Houseman pointed out:

$$f = \exp(i\alpha A)B\exp(-i\alpha A)$$
$$\Rightarrow \frac{\partial f}{\partial \alpha} = i\exp(i\alpha A)AB\exp(-i\alpha A) - i\exp(i\alpha A)BA\exp(-i\alpha A)$$
$$= i\exp(i\alpha A)[A, B]\exp(-i\alpha A) = i\exp(i\alpha A)B\exp(-i\alpha A) = if$$

Integrating this expression gives $f = K\exp(i\alpha)$. Putting $\alpha = 0$, we find $K = B$.

Prob. 2.1
Watch for exceptions! Any exception to the group rules means that the set and operation are <u>not</u> a group.

Prob. 2.3

This is best proved using matrix notation for boosts, determining the product of two boosts and using the symmetry – this is what I intend students to do.

Elisa Picarro pointed out that the group nature can be trivially understood using the cosh-sinh representation for boosts, or equivalently considering rotations (SO(2)) in Minkowski space.

Prob. 2.5

The matrix for a small rotation through an angle α about the z-axis is:

$$R_z = \begin{pmatrix} \cos\alpha & \sin\alpha & 0 \\ -\sin\alpha & \cos\alpha & 0 \\ 0 & 0 & 1 \end{pmatrix} \approx 1 + \alpha \begin{pmatrix} 0 & 1 & 0 \\ -1 & 0 & 0 \\ 0 & 0 & 0 \end{pmatrix} \equiv 1 + i\alpha L_z$$

So the generators are:

$$L_z = \begin{pmatrix} 0 & -i & 0 \\ i & 0 & 0 \\ 0 & 0 & 0 \end{pmatrix}, \quad L_x = \begin{pmatrix} 0 & 0 & 0 \\ 0 & 0 & -i \\ 0 & i & 0 \end{pmatrix} \quad \text{and} \quad L_y = \begin{pmatrix} 0 & 0 & i \\ 0 & 0 & 0 \\ -i & 0 & 0 \end{pmatrix}$$

where L_x and L_y are obtained from symmetry. It is easy to show that:

$$[L_x, L_y] = iL_z$$

obviously:

$$[L_y, L_x] = -iL_z \quad \text{and} \quad [L_x, L_x] = 0$$

and everything else follows from cyclic symmetry. So we find the Lie algebra:

$$[L_i, L_j] = i\varepsilon_{ijk} L_k$$

The structure constants for SO(3) are the components of the Levi–Civita tensor and are identical to those for SU(2) (Chapter 4), hence the two groups are isomorphic.

Prob. 2.6

The sum of the squares of the SO(3) generators are 2 times the identity. Since $2 = 1 \times (1+1)$, we see the correspondence with spin 1 ($j = 1$). Equivalently, since the generators for SO(3) correspond to $-i$ times the Levi–Civita tensor, the same result can be found using the identity:

$$\varepsilon_{ija}\, \varepsilon_{ijb} = 2\delta_{ab}$$

Prob. 2.7

The components of the adjoint have to be (painstakingly) obtained from the structure constants:

$$T^1{}_{23} = -if_{123} = -i\varepsilon_{123} = -i \quad \text{etc.}$$

We find $T^1 = L_x$ etc., so the adjoint matrices are identical to the generators and hence will obviously satisfy the same Lie algebra.

Prob. 4.1

Firstly consider:

$$J_3 J_+ \,|\, jm > \quad \text{and} \quad J_+ J_3 \,|\, jm >$$

Then consider:

$$J_- J_+ \,|\, jm > \quad \text{and} \quad J_+ J_- \,|\, jm >$$

General arguments about the generators follow, since the states $|\, jm >$ form a basis.

Prob. 6.1

This is *so* trivial, that it requires *no* algebra, but you have to spot the trick! And the trick is to consider the number of ways of listing p boxes with $(n-1)$ transitions of state label.

Prob. 6.2
Draw a generalised tableau and put in the numbers required to evaluate the numerator and denominator. Look at the transitions associated with the ends of rows and note the missing numbers in the sequences.

Prob. 7.2
Q only operates on the flavour states and σ_3 only operates on the spin states. The charge-flavour parts and spin parts of the calculation factorise. Work with the flavour ($\phi_{M,S}$ and $\phi_{M,A}$) and spin ($\chi_{M,S}$ and $\chi_{M,A}$) wavefunctions and avoid evaluating expressions until really necessary. Use the symmetries identified in Prob. 7.1.

Prob. 7.3
If we consider the first quark (quark a), the matrix element for the proton is:

$$< p \,|\, Q^a J_3^a \,|\, uud \, \frac{1}{\sqrt{6}} (2 \uparrow\uparrow\downarrow - (\uparrow\downarrow + \downarrow\uparrow)\uparrow) >$$

$$= < p \,|\, q_u uud \cdot \frac{1}{2} \frac{1}{\sqrt{6}} (2 \uparrow\uparrow\downarrow - (\uparrow\downarrow - \downarrow\uparrow)\uparrow) >$$

$$= \frac{1}{2} \cdot \frac{1}{6} q_u (4+1-1) = \frac{1}{3} q_u$$

For the 2nd quark (quark b), because of the symmetry between the first and second quarks, the matrix element is the same. For the 3rd quark (quark c), the matrix element for the proton is:

$$< p \,|\, Q^c J_3^c \,|\, uud \, \frac{1}{\sqrt{6}} (2 \uparrow\uparrow\downarrow - (\uparrow\downarrow + \downarrow\uparrow)\uparrow) >$$

$$= < p \,|\, q_d uud \cdot \frac{1}{2} \frac{1}{\sqrt{6}} (-2 \uparrow\uparrow\downarrow - (\uparrow\downarrow + \downarrow\uparrow)\uparrow) >$$

$$\frac{1}{2} \cdot \frac{1}{6} q_d (-4+1+1) = \frac{-1}{6} q_d$$

So the matrix element for the proton magnetic moment is:

$$< p \mid M \mid p > = \tfrac{1}{3} q_u + \tfrac{1}{3} q_u + \tfrac{-1}{6} q_d = \tfrac{2}{3} q_u - \tfrac{1}{6} q_d$$

From isospin symmetry ($u \leftrightarrow d$), the neutron magnetic moment is:

$$< n \mid M \mid n > = \tfrac{2}{3} q_d - \tfrac{1}{6} q_u$$

You can work out the magnetic moment of the Λ for yourself; but note that the u and d quarks appear as a $J = 0$ singlet and therefore their spins do not have a well defined direction with respect to the hyperon spin and cannot contribute to the magnetic moment.

Bibliography

[Barnes] Barnes, K. (2010), *Group Theory for the Standard Model and Beyond* (CRC Press).

[Close] Close, F. (1979), *An Introduction to Quarks and Partons* (Academic Press).

[Georgi] Georgi, H. (1999), *Lie Algebras in Particle Physics* (Westview Press).

[Halzen & Martin] Halzen, F. and Martin, A. (1984), *Quarks and Leptons* (Wilely).

[Lichtenberg] Lichtenberg, D. (1978), *Unitary Symmetry and Elementary Particles* (Academic Press).

[Merzbacher] Merzbacher, E. (1970), *Quantum Mechanics* (Wiley).

[PDG] Particle Data Group, *Review of Particle Physics*, published biennially in *Physics Letters B*.

Index

Baryon, 83, 90, 121
Baryon number, 49, 135

Cartan subalgebra, 81, 100
Casimir operator, 54
Chirality, 70
Classical mechanics, 16
Clebsch–Gordon coefficients, 84
Colour, 89
Conjugate state, 71
Conservation law, 14

Deuteron, 69

Flavour, 101

Gell-Mann matrices, 87
Generator, 10, 36
    rotation, 12
    SU(2), 53
    SU(3), 87
    SU(n), 43
    translation, 10
    U(n), 44
Gluon, 89
Goldstone boson, 137
G-parity, 121
Grand Unified Theory, 50, 134
Group, 25
    continuous, 29, 35
    cyclic, 27

exceptional, 142
finite, 27
Lie, 35
non-Abelian, 65
O(n), 43
S_2, 28
S_3, 28
SO(10), 142
SO(2), 29
SO(n), 43
SU(2), 53
SU(3), 87
SU(5), 50, 139
SU(6), 126
SU(n), 42
subgroup, 28
symmetry, 28
U(1), 29, 47
U(n), 41
Z_2, 28, 33
Z_3, 27, 31, 33

Hamiltonian
    classical, 17
Helicity, 21
Higgs mechanism, 136
Hypercharge, 101

Isometry, 43
Isomorphism, 28

Isospin, 101
  hadronic, 68
  quark, 69
  weak, 70

Jacobi identity, 40

Lagrangian
  classical, 16
Lepton number, 49, 135
Leptoquark, 141
Lie algebra, 39
Lorentz boost, 2, 15

Magnetic moment, 129, 130
Mass
  fermion, 137
  hadron, 132
  meson, 130
  W, 137
Meson, 81, 82, 92, 118
  pseudoscalar, 120
  vector, 120
Mixing angle, 120
Multiplet, 33, 103
Multiplicity, 105, 111

Neutrino, 134
  Majorana, 134, 141
  oscillations, 50
  sterile, 135

Observable
  conserved, 14
Operator
  antisymmetrising, 107
  exponentiation, 10
  Hermitian, 10
  symmetrising, 107

Parity, 2, 19, 33
Poisson bracket, 18

QCD, 89
Quark
  heavy, 129

Representation, 29
  adjoint, 39, 65
  fundamental, 59, 67
  irreducible, 33, 106
  reducible, 32
  regular, 31
Root, 81
Rotation. See Transformation
Rotation matrix, 62
  Spin 0, 62
  Spin ½, 63
  Spin 1, 64

Scattering amplitude, 95
Standard Model, 49, 132
  Lagrangian, 48, 135
Structure constants, 39
Superstring, 142
SUSY, 139, 142
Symmetry, 1, 13
  exchange, 106
  global, 48
  local, 51

Time
  reversal, 2, 15
  translation, 12
Transformation, 1
  gauge, 51, 65
  rotation, 2, 12, 30, 59
  translation, 2, 10
  unitary, 8, 29
Translation. See Transformation

Weight, 81, 101, 102

Young tableaux, 108